Praise  for

Ink

Finalist for the Governor General's Literary Award for Non-Fiction

"The history of ink and pens and the entire culture of writing by hand is a fantastic and (you knew this was coming) indelible subject for a book. But in the hands of Ted Bishop, one of Canada's best and most entertaining writers, the subject becomes a thing of rare beauty—and, best of all, a story you won't be able to put down. A brilliant accomplishment." —Ian Brown, RBC Taylor Prize–winning author of *The Boy in the Moon*

"Part cultural history, part memoir, and part travelogue . . . Bishop's writing is clear, researched, unpretentious, and moves along with a gentle humour." —*The Globe and Mail*

"The hugely talented Ted Bishop offers a spellbinding, sophisticated, surprising and deeply satisfying exploration of ink—making ink, writing in cursive, calligraphy or print, Islam's elevation of ink to the spiritual realm, and finally, wistful reflections on the renaissance of ink." —Elizabeth Abbott, author of *Sugar* and *A History of Marriage*

"In this latest quixotic, erudite, and entertaining adventure, Ted Bishop ambles from Budapest to Buenos Aires, and from Samarkand to Istanbul and New York City . . . Bishop transforms the conventional travel memoir into an unpredictable narrative of quirky intelligence and good humour. More, please." —Ken McGoogan, award-winning author of *Fatal Passage*

"A lively book. From start to finish, Bishop has cast ink's history in a strongly narrative mode, relying more on interviews with the living than plowing through books in the library . . . Bishop demonstrates impressive reporting skills throughout the narrative, as well as wit and an ability to describe scenic details." —*National Post*

"[Bishop is] an engaging and witty storyteller . . . *Ink* is both highly personal and exceedingly relatable." —*Toronto Star*

"Part travel narrative, part hidden history, part cultural exploration . . . *Ink* is a fascinating book, with writing as tactile and fluid as ink rolling across rice paper." —Will Ferguson, Scotiabank Giller Prize–winning author of *419*

"Ted Bishop reveals how the act of writing represents not just a flow of ideas but the condensed expression of a culture. His constantly surprising look at the craft, art, and spirit of ink is both quirky and humane, both inventive and wise." —Mark Abley, author of *Spoken Here: Travels Among Threatened Languages* and *The Prodigal Tongue: Dispatches from the Future of English*

"With his disarming voice, Bishop . . . proves to be an adventurous narrator who shares this story-enriched history with opening chapters about pen-and-ink inventions and discoveries, then of missing persons and murders, trade wars and financial double dealings, betrayals and unexpected friendships, soon to be intermixed with more and more of the author's own personal adventures and delightful insights— all somehow involving ink of every kind and viscosity." —Wayson Choy, *Literary Review of Canada*

Ink

TED BISHOP's first book, *Riding with Rilke: Reflections on Motorcycles and Books*, was a Canadian bestseller, a finalist for the Governor General's Award for non-fiction, and named a Best Book by CBC's *Talking Books* and *Playboy* magazine. He is a professor of English literature at the University of Alberta and writes with a fountain pen.

ALSO BY TED BISHOP

Riding with Rilke:
Reflections on Motorcycles and Books

Ted Bishop

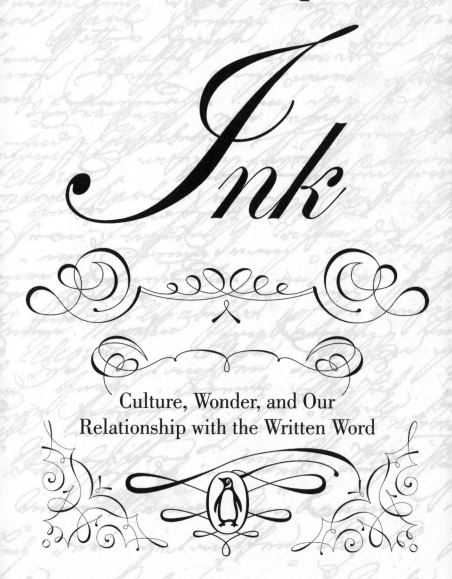

Ink

Culture, Wonder, and Our
Relationship with the Written Word

PENGUIN

an imprint of Penguin Canada, a division of Penguin Random House Canada Limited

Penguin Canada,
320 Front Street West, Suite 1400, Toronto, Ontario M5V 3B6, Canada

First published in Viking hardcover by Penguin Canada Books Inc., 2014

Published in this edition, 2017

1 2 3 4 5 6 7 8 9 10

Cover design: Lisa Jager
Cover images: Flourishes © Pyotr Bushuev; Letter © Roman Pavlik;
both from Shutterstock.com

Manufactured in the U.S.A.

Library and Archives Canada Cataloguing in Publication available upon request.

ISBN 978-0-14-316957-4
eBook ISBN 978-0-7352-3495-6

www.penguinrandomhouse.ca

Penguin
Random House
PENGUIN CANADA

For Thomas and the twins,
Nicola and Chloe,
almost old enough for ink

Contents

Introduction

Ink binds us. We are surrounded by ink, immersed in ink, a substance so common it is invisible. From cave walls to quill pens to laser printers, ink has traced the line of our culture. For millennia it was our social medium, and writing in ink used to mark our entry into the adult world. It was a rite of passage as memorable as that first drink, first drive, first kiss. But now children keyboard in kindergarten, pixels have replaced pigment, and pens will soon be as quaint as pocket watches. Even our signatures are electronic. Are we at the end of ink? This is the question I set out to explore.

We don't really see a technology until we're moving beyond it—the steam engine, the sailboat—but what was once utilitarian returns as a leisure activity and a luxury, like horseback riding, or taking the train, or using candles instead of electric light for dinner parties. As an English professor I'd spent my life surrounded by print, and hadn't concerned myself much with the material book until rumours of its demise began to circulate. As for ink, it hardly figured in the equation, as long as it was readable.

A writer at Edmonton's nonfiction literary festival remarked,

"The theme of all narrative nonfiction is the Quest. Playwrights write their endings first. With nonfiction you don't know how the book will end; you're writing to find that out." That may not be a universal truth, but it proved so for me. My quest began in a rare book library. I had some question about ink (now forgotten) that my sources on printing didn't answer, so I asked the curator, "What is the book on ink?" She said there wasn't one. I immediately decided to write a crisp little "commodity biography," with the required one-word-title-plus-elaboration (*INK: The Fluid That Changed the World!*); I'd wrap it up in a year. Writers notoriously dream up whole books in half an hour that take them the rest of their lives to complete. My one year turned to five as the clean line of ink I'd conceived became an ever-expanding blob, drawing me further and further into the social life of ink.

The project became a pilgrimage, taking me to Budapest and Buenos Aires, the home of Lazlo Bíró, inventor of the ballpoint pen. To China's Anhui province, where there's a factory that still makes inksticks from Ming dynasty patterns. To the border of Tibet and the world's oldest Buddhist print shop. To Samarkand to see the first Qur'an, soaked with the blood of the caliph who was assassinated for creating it. On the road I met characters whose stories threatened to take over the book: Bíró's daughter, who hid students during Argentina's Dirty War; Timor, the Muslim guide who washed his pork chops down with vodka; Mr. Chi, the earnest grad student who tried to save me from a drunken lunch at a Chinese ink factory; the two Steves, one who made me grind sheep bones in Texas, the other who taught me how to flame linseed oil in the Utah desert; Aung and Cheng, the two calligraphers who mocked and revered ink; Nathan, the "Willy Wonka" of ink, who railed against the government and produces inks that melt fountain pens. Closer to home,

I discovered a great-great-uncle who'd been a printer in the California Gold Rush, learned how my mother-in-law resembles Ming emperors, proved myself a failure at calligraphy, inked type, crushed gallnuts, and ground inksticks. Friends began to avoid me at dinner parties. I joined the FPN (Fountain Pen Network); I was invited to present at Nerd Nite. I saw ink everywhere.

I'm convinced that even scientific projects, no matter how arcane, from thermodynamics to string theory, are rooted in the personal, even if the writer isn't conscious that it is so. I never knew my grandfather, Edward Thomas Bishop, and aside from a graduation photo from Osgoode Hall I have no idea what he looked like, but I inherited his library. There are no letters, no diary, but he signed all his books, and while the stiff photo holds the viewer at a distance, the ink on the page always made me feel a connection. He exists for me as a signature.

In my professional life, the experience that's had the greatest impact was an encounter with ink. As I wrote in *Riding with Rilke*, I was a young Virginia Woolf scholar working in the British Museum Library, reading her correspondence:

> I opened up the next manila envelope and slid out a single sheet. I found myself reading a letter I had read in print dozens of times before. Anybody who works on Woolf practically knows it by heart, it's reprinted so often. It begins:
>
> > *Dearest, I want to tell you that you have given me complete happiness. No one could have done more than you have done. Please believe that. But I know that I shall never get over this: and I am wasting your life. It is this madness ...*

I felt a physical shock. I was holding Virginia Woolf's suicide note. I lost any bodily sense, felt I was spinning into a vortex, a connection that collapsed the intervening decades. This note wasn't a record of an event—this was the event itself.

I turned the sheet over.

There Leonard had written in green ink the date: 11/5/41. This detail set off an unexpected aftershock. I had seldom thought of him, of how he had had to wait twenty-one days before the body was found. Three long weeks, answering questions from *The Times*, taking calls from friends. Then a group of teenagers, throwing rocks at a log in the river, found it was not a log at all and dragged what was once Virginia Woolf ashore.

The episode taught me about the impact of the material text, but what I hadn't considered was that what I was responding to, as with my grandfather's signature, was the ink.

I'd come to find ink itself compelling: the *slurp slurp* in the print shop as the thick goop warms on the rollers, becoming ready for the type before emerging on paper as deep black letters; the infinite grey shades of Chinese calligraphy; the way it flows from a well-tuned fountain pen. I talked to print-culture historian Michael Winship, who said, "There is a mystery, a magic, to ink—how does it know when to be liquid, when to be solid?"

Travel and research only really become interesting when things fall apart, the neat itineraries go by the wayside, one thing leads to another, and you're led off the map. The ballpoint, which I thought to dispense with in a paragraph, proved fascinating. After half a century of development it still demands the most exacting technology, yet it sells for less than a cup of coffee. It is abused and ignored, and seems to have become prominent in the popular imagination only as a weapon (Joe Pesci kills a rude thug in *Casino*; Rachel McAdams stabs the creepy Cillian Murphy in the windpipe in *Red Eye*; The Joker in the Batman movies goes for both the throat and the eye; and Matt Damon disables an assassin with a Bic in *The Bourne Identity*). In fact, the ballpoint found its first success as an adjunct to a weapon—it was sold to Royal Air Force bomber crews because it worked at high altitudes. With Chinese inksticks I discovered that ink could be an exquisite object, the subject of poetry, a means of entry to the emperor's court, and a thing that engendered betrayal, murder, and the theft of concubines. The ink of the Qur'an led me to recipes for ink with anemones, the rind of pomegranates, and the gall of a turtle; to the union of the erotic and the sacred in Arabic letters; and to the Sufi mysticism of a Canadian poet. It was a haphazard line from a ballpoint on my desk to the ink cakes at the Met in New York to the illuminated Qur'ans in Istanbul. Only later did I begin to see a progression, from the utilitarian to the aesthetic to the religious engagements of ink, reflected in the sections of this book: Craft, Art, and Spirit.

The basic elements of ink are simple—pigment, liquid, and a binder to make it stick—but its applications are endless. I came away convinced that the need to write is primal, and is linked to a desire for permanence that has recently become more urgent. You too, I'm sure, have faced the Blue Screen of Death on your computer, had your documents go—where? Not up in smoke, or ripped or rotted, but into the no-where of cyberspace. Not even ash or shard to testify they had once existed. But you too have had the sensuous pleasure of a handwritten note, and maybe you too, though self-conscious about your half-printed, half-cursive scrawl, have returned to the pen, chosen your ink, discovered that it is a medium that still carries a message.

Ink

PART I

THE *Craft* OF

Just as much a boon for democracy as the transistor radio or the quartz clock. The bastions of privilege based on money have fallen.

—LAZLO BÍRÓ, ON THE SIGNIFICANCE OF THE BALLPOINT PEN

THE BALLPOINT

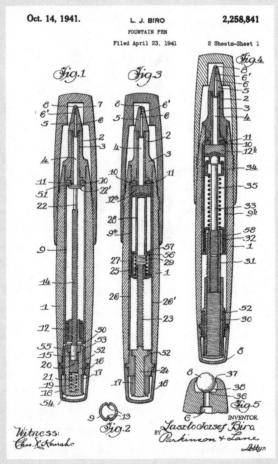

The drawings for Bíró's ballpoint pen patent aroused the suspicions
of a Spanish border guard, who thought they were plans for a torpedo.
1941 U.S. patent, image courtesy of the author.

Desperately Seeking Bíró

I

On October 29, 1945, hundreds of people lined up outside Gimbels department store in New York City, so many that fifty policemen were on hand to keep them in line. They were there for the "miracle pen" that would "revolutionize writing." The excitement was like that for the iPhone, and for the people there the technological advances were as amazing. A pen that carried its own inkwell! Unheard of. They were expensive at $12.50—the equivalent of $130 today—but they were worth it. They would write for six years! Fifteen years! They would write underwater!

And this pen—it wasn't a Bic, it wasn't a Parker, it was the Reynolds Rocket, the brainchild of Chicago entrepreneur Milton Reynolds, who had thrown it together in four months. An extravagant promoter, he completed a round-the-world flight in a converted bomber, beating Howard Hughes's record, handing out pens at every stop. He made an unauthorized flight in China to try to find a mountain higher than Everest (which he would name after himself), and when they tried to arrest him he threw cartons of pens at the Tommy-gun-toting guards, told the pilot to gun it, and escaped to Japan. He presented a set of two hundred pens to the president, engraved with the inscription *I stole this pen from Harry S. Truman.* He was in the press every week.

Today, no one remembers Reynolds because most of the pens didn't work. They didn't write for six years, sometimes they didn't even write for six hours before they quit or leaked all over your shirt. Outraged customers demanded replacements, refunds, and dry cleaning. Soon the price had dropped to $3.85. Two years after their introduction they were selling for 19¢ and nobody was buying. The ballpoint was dead.

But of course they're not dead. You're using them. They are simply so ubiquitous that they're invisible. Where did they come from?

The ballpoint pen was invented in 1938 in Hungary by Ladislao "Lazlo" Josef Bíró (pronounced BEE-roh), sports-car racer, amateur inventor, and journalist. He fled Hungary because of its anti-Jewish laws, and in 1943 patented his invention in Argentina. There he met a British banker, Henry George Martin, who bought the rights and introduced the pen to the Royal Air Force, which had been seeking an instrument that could write at high altitude for the navigators of their bomber crews. They worked, and "Biro" became shorthand for ballpoint throughout the Commonwealth. But of the millions who use the pens (and mispronounce the name as "BYE-roh"), few have heard of the man. I decided to seek him out.

II

I was travelling with my wife, Hsing, who tolerated my ink mania, liked good fountain pens (which she felt should always be carried in Furla handbags), and loved cities. We took the EasyJet night flight from Geneva to Budapest. It was an uneasy ride—the plane was packed, the announcements were incomprehensible, and the guidebook warned of taxi rip-offs and muggings. In the terminal I veered away from the pretty girl handing out flyers for Zona taxi and stood by the minibus desk with the other apprehensive foreigners, memories of Cold War spy novels churning in my brain.

The driver in his thick leather jacket was straight out of John le Carré. He chucked our luggage in the van, gestured to the door, didn't meet our eyes. At the hotel he slung the bags onto the sidewalk like someone dumping a body and walked past us as if we didn't exist. No cheery "Welcome to Budapest!," no hovering for a tip. Was this the legacy of Stalinism? If you look at nobody, then when you're interrogated you can truthfully say you haven't seen anyone. The Fiesta Hotel had endless Day-Glo yellow-green hallways that made you feel you had a sick hangover by the time you got to your room. The room had a skylight but no windows; the coffee bar downstairs had no windows. The whole place felt like a prison that someone had tried to brighten up, and the desk clerk was as grim as a warden.

Or maybe not. I'd come looking for background on Lazlo Bíró, the journalist who invented the ballpoint pen. I knew the basics: annoyed by the way his fountain pen smeared proof sheets, Bíró watched the printing press laying down thick, smudge-free ink. It came to him in a flash—put that ink in a tube, insert a ball bearing at the end, and you'd have a mini rotary press that would generate just enough friction to warm the ink paste and then allow it to dry instantly. Brilliant. I figured finding Bíró would be easy, like James Joyce in Dublin—there would be Bíró statues, Bíró plaques, Bíró walking tours, Bíró T-shirts, maybe a Bíró pub, and certainly overpriced, made-in-China replica Biro pens. But no. Nothing.

In a bookstore on Váci, the main shopping street, I asked about books on Bíró. "No. We have nothing like that." Bíró was also a painter, his work housed in the state museum, so I inquired at the Museum of Fine Art. "No! We have no twentieth-century art here." They sent me back across Heroes' Square, under the scowls of its warrior statues again, to the Palace of Art. There they said they'd never heard of paintings by Bíró. "Try the Hungarian National Museum." I walked the half hour down to the old area of Pest. In the museum lobby there was a big circular desk with an *i* on it and a security guard leaning on the counter, flirting with one of the young women. The other young woman at the desk was friendly and apologized that this museum had no paintings. "Only things," she said. (They had Beethoven's piano but I didn't care, I wanted Bíró.) She said, "I will phone the National Gallery." She was transferred, transferred again, made notes, called another number, and reported that one section had said "Absolutely they had nothing by Bíró" and another had said "Maybe they did have something by Bíró in graphics." I should come over there and talk to them.

The Hungarian National Gallery—not to be confused, as I had done, with the Hungarian National Museum—was across the river and up on the hill. The river was the Danube, the hill was Castle Hill in Buda (Buda and Pest are like Minneapolis and St. Paul or Strathcona and Edmonton: once separated and bridgeless, with only ferries going between them, each superior to the other), and the gallery was in a wing of what once had been the Royal Palace.

The curator came down to speak to me. "Bíró? Yes, of course

we know of Bíró, but I think you will find nothing here. Not in Hungary. Maybe abroad—in Argentina, yes—not in Hungary. He left, he did not die here."

"But Bíró invented the ballpoint pen—and he was a well-known journalist, a medical student, a race car driver ..."

"Yes yes, it is sad. It is the same with Moholy-Nagy. He is not known here. In Germany he is famous because of the Bauhaus. But he left, he did not die here."

"I know, but Bíró didn't leave until he was thirty-nine, a full lifetime. He had already patented a washing machine, an electro-magnetic mail system, and an automatic transmission that he sold to General Motors ..."

"He left," she said. "He left. He did not die here."

I told her that in Canada we cheerfully claim anyone who's ever lived there, or stayed longer than twenty minutes. "If you change planes you qualify," I said brightly, all set to tell her about Mavis Gallant and Malcolm Lowry, but stopped. She understood the English but not the concept. In Hungary, you go, you're gone.

I started thinking of service staff as the Scowling Magyars. From curators to clerks to waiters and tram conductors, they all scowled. Scowled like the statues of the warriors with the fierce moustaches in Heroes' Square. Hsing read the guidebook. "It says the suicide rate is high here—it was the noble way out for bankrupt aristocrats and so didn't carry a stigma. You would still be buried in consecrated ground. Hanging is preferred."

"Hence the hangdog expression of all servers. They're just going to finish their shift and go home and hang themselves. They're already depressed because the rope will probably slip or the light fixture will come out and they'll have to try two or three times." I later read about a doctor who committed suicide by tying a rope to a chair and strangling himself on the floor. That's determination.

Hungarian seemed as grim. I figured out that *kaveház* was coffee house and *étterem* ("eat-a-room") was a restaurant, but although the word for "thank you" sounded vaguely like "kissin' him" it brought no response when I used it. In Italian the accent routinely falls on the second syllable, giving the language a lilt; in Hungarian the stress falls on the first syllable (*utca* = OOT-ka), which makes the language emphatic, like chopping wood. Hsing was much better, maybe because she was used to the tones of Mandarin and had done years of ear training for violin. She mastered the "thank you"—*köszönöm*—and quickly figured out that *utca* is a street, *út* a smaller road, and *körút* is a ring road. Also that *könyvtár* is a library, and so we made our way along various *utca*s and down *út*s to the Applied Arts Museum in search of Bíró.

In the museum I pushed through an oak door marked *Library*. The hinges squealed and the aged battleaxe at the coat check, obviously the mother of one of the stern warriors in Heroes' Square, scowled. I wondered what fierce crone would challenge me on the other side. But the librarians smiled, found a couple of books on fountain pens, and when they could do nothing more, gave me directions to the Technical University library. There the pattern of frosty reception and helpful engagement repeated itself. I asked if Bíró was well known in Hungary. "Oh yes," said the first librarian in a tone that implied "Who cares?," but the director found me a little pamphlet in English and told me I could copy it. I found the one old photocopier that cranked and whirred and spat out a page every thirty seconds, ministered to by a leathery-faced dwarf. Though she looked grim she proved only shy, and told me her English was "much better before." I wondered if I should ask "Before what?," but the director came back and told me there was material on Bíró at the National Széchényi Library, across the river in Buda, up on the hill in the palace.

I paid my six thousand forints, stood by the pillar, got my picture taken, got my card with a bar code. I was now not a tourist, I was a Reader—A 76280—and I learned how to drop the card in the slot and tip it forward so that the barcode scanner could read me, turn the little light green, and click the turnstile open. Lesser mortals stood on the other side and looked up at the broad marble staircase with the red carpet that led to the reading rooms.

The woman at the reference desk spoke English and helped me fill in the request slips. Yes, they had bound journals of the arts magazine Bíró worked for, *Hongrie-Magyarország-Hungary*, and microfilm of another magazine, *Elöre*, plus an English translation of a book about the ballpoint, *The Never Ending Line: The Ball-Point Legend* by György Moldova. This turned out to be the gem. Research is like mining: you sift through a lot of gravel hoping for that one nugget; or like detective work, where you sort through endless documents knowing that there'll be one that will crack the case.

Things were looking up. As we walked back through the university district, Hsing saw a door with a small sign for a rental agency. "Let's buzz it," she said. "Maybe we can get out of that horrible hotel." I was diffident—what if they replied in Hungarian?—but Hsing is forthright and will not tolerate bad lodgings. Tibor, the young manager, spoke excellent English, and in minutes booked us into an apartment with soaring ceilings and parquet floors. In the nineteenth century it would have belonged to a wealthy bourgeois. I was glad to make the shift from Stalinist modern. I paced off the apartment: twenty-nine strides from the bedroom at the back to the tall window overlooking the turreted Science Library with its warm lemon-yellow walls. I checked the guidebook. "Founded in 1561 by the Jesuits. Two million books. I want to live here," I said.

III

Back in the library the next day, I learned that Lazlo was the wild child of the family. His older brother, Georg, went to university and like his father became a dentist; Lazlo studied medicine for a few semesters but was only really interested in hypnosis, and dropped out of medical school. Bíró claimed to have been, among many other things, a biological researcher and a racing driver, but apparently the research consisted of dissecting a couple of frogs and the auto racing was a one-time event.

Bíró bought a red Bugatti sports car from an actor named Svetislav Petrovic after a night of drinking. To put this in perspective, it would be like hanging out with, and at the end of the evening buying a racing car from, Steve McQueen, Hollywood's most famous auto aficionado (who was of partial Hungarian descent). Petrovic started out in Hungarian silent films and would go on to make more than a hundred over the course of his career, including an appearance in a 1958 Louis Malle thriller with Jeanne Moreau. At the time he sold Bíró the car, Petrovic had already appeared in more than thirty films. The only thing more glamorous than the chisel-jawed actor would have been the Bugatti itself, more exclusive and costly than a Ferrari today. No Bugattis of that era had twelve-cylinder engines—but the outrageous Bugatti Royale had a massive twelve-litre motor (think two and a half Hummers). I had no idea how a journalist could have financed it; perhaps alcohol was a factor. Bíró does say that he soon sold the car because he couldn't afford it.

It was under the influence of champagne that Bíró offered to enter the car in a race on Sváb hill in Buda two weeks later, though he had never driven. He took lessons from an old chauffeur friend, but would often forget to put in the clutch when he shifted, grinding the gears. The chauffeur would punish him by stomping on his foot. Bíró claims he won the race, but that wasn't what made the event significant. It was his difficulty in working the clutch that gave him the idea of the automatic gearbox.

He worked for a year with an engineer to create an automatic transmission. They installed it in a 350cc motorcycle, and, with the two-hundred-pound engineer in the sidecar, drove all the way from Budapest to Berlin where they met with representatives from General Motors. GM was impressed and signed a contract with Bíró for half a percent of the sale of units produced and an advance of $200 a month for five years, half for Bíró, half for the engineer. This was considerable cash in 1932, and so they eagerly signed the contract, without consulting a lawyer. Only later did they see that GM had made no mention of when they would begin production, and how many gearboxes would be made. In fact GM had been working on their own automatic transmission for years, and had signed the contract with Bíró only to keep his invention off the market. The same combination of brilliance, exuberance, and naive trust would mark Bíró's dealings with the ballpoint pen.

Inventing ran in the family: Lazlo's father invented something called the "rubber porter" that allowed the front door of the apartment to be opened from any room, and experimented with a new kind of pen. Instead of being filled with ink the pen was filled with water, which flowed through a thin tube and dissolved an ink-cartridge core. The pen didn't work because it couldn't supply an even flow of ink—exactly the problem that bedevilled Bíró for years with his own pen. I was beginning to see that Bíró had a journalist's nose for the powerful anecdote: expanding on an idea your dad had been fiddling with since you were a kid is much less striking than a bolt of insight in a print shop. In fact Bíró's first patented invention was a "water fountain-pen" in 1928 (he was born in 1899), though we know nothing about it.

Moldova referred to *Hongrie-Magyarország-Hungary* as a "fine-art magazine," so I ordered up the library's few back issues. I undid the binder twine, expecting an avant-garde art periodical. The title was trilingual, the first page laid out in columns of French, Hungarian, and English, but it was not at all avant-garde. The lead story, "Public Hygiene and Welfare Institutions of Budapest," proudly announced that lifeboats on the Danube had saved 189 lives out of 194 boating accidents the previous year (an accident every second day; the citizens of Budapest must have been good swimmers). An article followed on "The Development of Hungarian Furrier Industry." Uninspiring stuff, and although Bíró was listed as managing editor, his byline appeared on none of the articles. Editing *Hongrie-Magyarország-Hungary* must have been simple work for him, a few hours sifting through bland articles over coffee while his creative mind worked on his inventions. In the second issue I found the statement of purpose: "To give a picture of our country to the Western States." This was just a promotional trade magazine, and by 1934 it was bent on assuring the English and French that Hungary would be a good ally against Germany. In response to doubts voiced by a French politician, the editors of *Hongrie-Magyarország-Hungary* exclaimed: "If they only read our history! Where were the Czechs, Serbs and Rumanians when we fought these wars? They fought against us, on the side of the Germans!"

I wondered about Moldova. In researching his book he'd used Bíró's memoir, *Una Revolucion Silenciosa*. Not only was Bíró's biography written in Spanish, dictated to Argentine journalist Hector Zimmerman, but Bíró cheerfully advises his readers that they "should never forget that what they hold in their hands is a hopelessly biased work. What I recount is the truth, but it is probable that the facts and persons presented are not in reality quite as I describe them." Moldova also relied on Andor Goy's unpublished memoir, truculently titled "The Real Story," written, Moldova admitted, when Goy was "losing his intellectual powers." In it Goy, Bíró's early partner in the ballpoint, rages against Bíró's memoir: "I could not find one honest sentence in it to accompany the flurry of self-praise. It is arguably not worthy of being put through a printing press." He also accuses Bíró of making elementary mistakes in grammar and spelling—usually the last refuge of critics who don't like the content.

Moldova himself decided to dramatize the story, asking the reader to consider his book "both a historical manuscript and a novel." So, a creative nonfiction composed of creative nonfictions. Further, what Moldova delivers is the clash of two Hungarians, not the story of the ballpoint pen: Marcel Bich, who perfected the Bic pen, is never mentioned, nor is the globe-girdling flim-flam promoter Milton Reynolds, who created the market. And Moldova's book, indispensable as it is, was a vanity commission: published by ICO, the Hungarian pen company, to mark their fiftieth anniversary. Bíró was taking shape through a veil of semi-fictions. Maybe that's all there ever are.

The window was open to the Buda hills, traffic hum floating up from below, punctuated at regular intervals by the screech of a tram. The library was closing. Bíró's anecdotes were his true made-up stories, his true creative embellishments of memory. Who was I to quibble about whether a Bugatti was twelve litres or twelve cylinders? It was huge, it was fast. Wasn't that enough?

IV

At the base of the hill below the library, beside a busy traffic circle, there stands a vulva-like ellipsoid in white concrete with *KM* inscribed on the base. Who is KM? Hsing found it in the guide-book: KiloMetre. The statue, not meant to be erotic at all, is the zero kilometre mark: all distances from Hungary are measured from here. This was the epicentre. But only for road makers. After my day at the library Hsing dragged me back to Liszt Ferenc tér (Franz Liszt Square), to the Liszt Academy to buy tickets for some pianist that night. This Liszt Academy looked like serious business: a heavy-pillared facade with a green statue of Liszt on a throne, his plinth held up by naked, muscle-bound workers. At the top women with jutting breasts brandished laurel wreaths, just in case you were wondering if Liszt was the man or not. I was grumpy; I wanted to hear some gypsy jazz at the café near our place.

"Hey," said Hsing. "This guy is world famous."

"Not to me he isn't, and his name—'pariah'—doesn't sound promising."

"Look, he's like the Mick Jagger of pianists, okay? You'll be glad you went. You just need something to eat." I appreciated her attempt at an analogy for my generation, and I know that when my blood sugar drops I get "hangry," as she puts it, so we stopped at a sidewalk café near the Academy. Hsing had tea and read about how the vivacious Hungarian Zsa Zsa Gabor had an affair with Atatürk, the founder of modern Turkey ("I never accept gifts from perfect strangers," she said. "But then nobody's perfect."). I ordered the crispy leg of goose on pasta with sauerkraut, and a Hungarian beer, Dreher Pilsner. It was terrific, as was the piano guy. His name was Murray Perahia, not "pariah," and the place was packed. He played standard hits—a Bach Partita, Beethoven's "Pastorale," Brahms, and six short pieces by Chopin—which suited me well and which Hsing pronounced fabulous, though she'd hoped for some Liszt or Bartók. At the end the audience clapped in unison as they do at hockey games and brought Perahia back for bow after bow. "He won't do an encore, he's tired," said Hsing, shouting in my ear over the roaring crowd. These fierce Hungarians take their art seriously. As we tumbled out into the square I realized that this was the epicentre—the Academy, the cafés, the space itself an incubator of genius. Would Bíró have come up with his ideas in a suburb? The place seethed with creativity.

"What do you mean, 'seethed'? You didn't even want to come."

"You know what I mean."

*Bíró invented many devices, including a cigarette holder. He smoked
a hundred cigarettes a day—"But," he said, "I drink a lot of coffee."
© Fundación Bíró, courtesy of Mariana Bíró.*

Bíró in Budapest

That city of cafés and newspaper addicts ...
—FRIGYES KARINTHY

I

In 1900 there were over five hundred coffee houses in Budapest—more and better than Vienna's, according to Budapestians, who are always comparing themselves to Vienna. At the Bristol Café young Bíró could have run into musicians Béla Bartók and Zoltan Kodaly, literary critic Georg Lukács, sociologist Karl Mannheim, photographer and Bauhaus designer Lazlo Moholy-Nagy. Bíró's office was a ten-minute walk from the fashionable Japan Café on Liszt Ferenc tér. Writer Ernő Szép makes the café sound magical: "Its walls [were] covered with majolica tiles painted over with bamboos, chrysanthemums, vases and dream-like birds ... the café was a more distantly exotic place than the world of white lotus, green tea and the golden Buddha." A place not just for pastries and coffee but of and for the imagination: "the Japan was the fairy land of youth. That was where I went in the afternoons, ignoring novels, horse-races, and even love sometimes; such sacred need of life that was in the time of amicable fraternisation." The Japan Café not only featured exotic tiles to feed the imagination; it carried hundreds of periodicals in various languages, and kept ink, blotters, and long strips of paper (called "dog's tongues") in the basement for the patrons—this was the café of journalists and writers. Bíró spent more time working here than at his office, and it was here that he had his flash of insight that changed the way the twentieth century would write.

In his memoir, *Una Revolucion Silenciosa*, Bíró tells of how as a journalist he was frustrated with his Pelikan fountain pen—it was always running out of ink or drying up, and the nib split if he pressed on it. He envied the way the rotary press spread the ink on the page, and the way the ink dried, smudgeless, the moment it touched the paper. It was in the "loud mechanical confusion" of the print shop that the idea of the pen with printers' ink came to him. But he had no idea how to make such a thing. He wanted something better than the fountain pen or the typewriter, something with rollers like the printing press, except that the press rollers only turned one way. He came back to the Japan Café to work on an article, still absorbed in the idea of a pen that worked like a press.

"The usual?" asked the waiter, and Bíró, still mulling over the rollers, ordered his coffee not with a splash of milk but with a "splash of ball," and with that slip of the tongue realized that what he needed for his pen was not a roller but a ball. The ballpoint was born. Or at least conceived.

In later interviews Bíró would return to this moment in the Japan Café. He resented those journalists who made up stories— that the idea was born when a few little balls on Bíró's desk rolled through some spilled ink and drew lines in front of his eyes, or that he was sitting at a café watching kids play marbles and noticed that after passing through a puddle the marbles left a trail on the tarmac. Bíró gives the story pride of place in *Una Revolucion Silenciosa*, opening with the waiter's query, "Que va a tomar el señor redactor?" ("What would you like, Señor Editor?"), and his reply, "… ¡con bolilla!" ("With a ball").

It's a "Eureka!" moment, like Archimedes leaping from the bathtub, having suddenly discovered the principle that a floating object displaces its own weight of fluid, and running naked into the street. A great story and just as unlikely. Bíró insists, "What I was looking for was something that was entirely new among writing instruments. I knew the origins and development of writing tools in considerable detail, from symbols inscribed in stone … to the fountain pen, but none of these precedents provided me with any kind of guidance with my work." Moldova points out that Bíró forgets to mention the American patent of 1888 by leather-worker John Loud, who devised a kind of roller ball to write on leather; Anton Sheaffer's German patent for a ballpoint, October 1901; Michael Braun's of October 1911; and the presentation at the 1924 Dresden trade fair of a finished ballpoint called the Mungo. In Hungary Dr. Dezsö Ránk and railway engineer Ödön Hajdú had patented their version of a ballpoint in 1934—and Bíró himself had a decade earlier patented something called the "water fountain-pen," which was probably a development of his father's idea. That flash of inspiration in the Japan Café? Born of a decade of thought.

II

Each morning I would go across the street to the Centrál Kavéház, where I worked for an hour over coffee before the library opened, thinking of Bíró's transit from café to office and back. Then I walked down Iranyi utca to the Danube, jumped on the back of the #2 tram, and rode the three stops along the Duna-korzó— past the moored tour boats, the restaurant boats, the jazz club boat *Columbus Jazzklub*—down to Roosevelt tér. I got off and walked across the Chain Bridge. I knew to keep to the right at the two kinks in the middle, to watch out for the cyclists who cut the corner coming through. In the first kink a man stood beside a low box with bright stones displayed on a blanket. No sign, no sales pitch; he simply stood and smoked, looking at the boats on the Danube as if he were just watching over the stand for a friend who'd stepped away for a moment. The weather was bright and the Danube the colour of Lake Louise. I knew the river was polluted but I loved it, especially the heavily laden barges that looked like they were going to be swamped by their own bow waves.

I passed the funicular railway (expensive, anticlimactic, and thronged with tourists) and marched up the stone stairs through the treed slope, breathing hard by the top but glad of the brief workout, to come out in the courtyard of the palace. I had new Hungarian shoes with stiff lugged soles. Cobblestones make your ankles wobble from side to side, and the stones in the courtyard had big gaps between them that threatened to pitch you right over. You go through the archway, around the back past a statue of hunters and a dead stag, to the esplanade. I always walked over to the edge and looked out toward the Buda hills. After I'd learned about Bíró's car race, I wondered if one of the ones I was looking at was Sváb hill.

At ten a.m. the guard unshackled the iron gate and everyone piled lunches and litre bottles of water on the window ledge by the lockers, a homey and trusting touch that contrasted with the barcode-operated turnstiles at the base of the stairs and the glass security doors upstairs. I sat in seat 658 at a Formica-and-steel table. The first day they told me, "You must wait one hour for the book, one half hour for the journal and the microfilm." Better than the British Library where sometimes you wait days. Now they had my books waiting for me at the desk.

When I first encountered Andor Goy's name in *World Famous Hungarians* I knew nothing about him, but a Hungarian design website listed the Go-Pen by Andor Goy among Hungarian inventions. No separate entry for Bíró. Moldova at least gave them equal billing. Their photos look out from the back of Moldova's book. Bíró is smiling, seems to be snapping his fingers; Goy is serious, even disapproving. Both men are middle-aged and balding, but you can see the kid in Bíró. And that Goy never was a kid; in his youthful pictures he already has the gravitas of middle age. I learned that Andor Goy's shop was on Roosevelt tér in the Gresham Palace (now the Four Seasons), on the Pest side of the Chain Bridge where I caught the tram. More to the point, I learned that Goy hated Bíró's guts. I could see why Bíró wasn't more celebrated: he had renounced his Hungarian citizenship to get his family into Argentina, while Goy stayed in Hungary and saw his company absorbed into the Hungarian national pen company. Bíró had become famous world-wide. Goy was famous only in Budapest, where he spent the rest of his life gnashing his teeth and insisting, "I'm the guy, not him." But aside from the Hungarian cultural website, nobody cared. Goy was prominent, his widow bitter; putting up a statue to Bíró would have been a slap in the face. And Andor himself made it a conflict of Goy versus Jew.

It was Imre Gellért, Bíró's friend since childhood and current financial backer, who brought in Andor Goy. Goy's real interest was the typewriter—he had a repair shop with over a hundred workers, and he was developing his own invention, the automatic return typewriter. Even so, in March 1938 he met with Gellért and Bíró at the Japan Café, bringing with him a Mungo ballpoint pen he'd bought at the Leipzig trade fair six years before. Bíró and Gellért were already there, their hands covered with ink, showing their new pen to the café regulars. In Moldova's telling of it, Bíró looked at the pen and said, "This really is a ball-point pen. Judging from the stains on it, it worked with ink.... The whole point of my invention is that it uses a special mixture of dye paste instead of ink. And that is adequate justification for a patent."

Design problems plagued them. At this point they weren't working with capillary action; the pen was like a syringe, with a piston to press down on the thick liquid, but it only wrote in the vertical. And if they applied too much pressure the ball would pop out and spray dye in all directions. The big problem, however, was the ink. Bíró writes, "If on one day the liquid appeared too thick, by the next day it would be too thin, and after a while it would become thick again. It was only much later that we became aware that the changes displayed by the dye material depended on its temperature.... Moreover the aniline would often crystallize, rendering the entire mixture unusable." He and his brother went to talk to a chemistry professor and explained that they were looking for a dye that would remain fluid in the cartridge but dry as soon as it touched the paper. The very notion irritated the professor: "'This is *contradictio in adiecto*, contradiction in the premise itself, like ice that is both hot and engraveable: it does not, and cannot, exist!'" Yet it would.

Despite the problems, Goy agreed to go into business with Bíró; the patent was registered six weeks later, at the end of April, and a month later, on May 30, 1938, Bíró and Goy signed a contract. The timeline is important. Two days before, on May 28, the Hungarian government had passed the first of a series of anti-Jewish laws. The new act decreed that professions such as medicine, the law, engineering, acting, and journalism could be practised only by members of the relevant professional association—and Jews could not make up more than twenty percent of any of those associations. In *Una Revolucion Silenciosa* Bíró tells of how he always bought cigarettes in the shop of an old widow whose husband had been honoured for bravery in the First World War. One day she told him that her shop licence had been revoked because a family member had married a Jewish girl. Bíró, thinking that the prestige of being a journalist for *Előre* would carry some weight, visited the relevant division, where the official turned out to be an old friend. The situation looked promising, until the official interrupted with, "You know very well, Mr. Bíró, that you yourself are one of the 'undesirable elements.'"

Goy took the pen to Germany, to the Faber-Castell factory in Nuremberg where the technical director said, "This crude instrument is no competition for a classical steel- or gold-nibbed fountain-pen"; to the Pelikan factory in Hanover where they turned him down; to Montblanc in Hamburg where he was again rejected; and finally to Munich's DFW Deutsche Füllhalterwerke, where, according to Moldova, Goy was greeted by a manager with a Nazi armband. They were interested. But there was a catch. As Bíró discovered, in his last meeting with Goy two days before Christmas in 1938, the agreement Goy signed with the DFW committed him "to help the creation of a Jew-free Europe by refusing to sell the products of Deutsche Füllhalterwerke GmBH, whether directly or indirectly, to any Jewish company."

III

I dropped by the apartment-booking office to settle my account and told Tibor about Bíró's dealings with Andor Goy. Bíró had said, "If I fail they will say I am Jewish. If I succeed they will say I am Hungarian." The manager said, "You know, Hungary has the highest number of Nobel Prize winners for its population. No one can figure out how this can be for such a small country, but—they are all Jewish!"

As of January 1, 1939, a new law would make it illegal for Hungarian intellectuals to leave the country. Bíró jumped on a train for Paris on New Year's Eve, 1938, leaving his wife and daughter behind. (A second anti-Jewish law in 1939 would prevent Jews from becoming newspaper editors, and a third in August 1941 would make marriage or sexual relations between Jews and non-Jews illegal.) In Paris he drank coffee with other Hungarians at the Café des Deux Pistolets, and worked on the ink. He also became involved with something more sinister: pressured to participate in the French war effort, Bíró developed a precursor to napalm. "I mixed potassium chlorate with carbon tetrachloride and white phosphor in the right measure, and sucked it up with wood-pulp," he notes. "Sealed in a thin-walled glass container, it would be suitable for use as an incendiary bomb to be dropped in large quantities from an airplane." But one of the cylinders blew up, splashing the liquid into his partner's face, so Bíró decided he wanted out of the project.

In Budapest he'd met General Augustin Justo, former president of Argentina, who offered him his card and told Bíró to contact him if he ever wanted to make the pens in Argentina. He also knew Maria Lang, a patient of his brother Georg, who was moving to Buenos Aires to be with her businessman husband and, intrigued by the ballpoint prototype, drew up a contract offering them financial backing in return for the rights to produce the pen in Argentina. It was time to go. From Barcelona he wrote to friends about a letter from Andor Goy that would haunt him for years: "According to

[Goy] my 'escape' was a typical example of cowardice, further evidence that I was a 'traitor'. My escape was completely in vain, as however far I got from them there would be no place on the Earth where they would not reach me. He made mention of the superiority of the Aryan race."

Bíró departed for Buenos Aires lonely for his wife and daughter, with a pen that still did not write and caused nothing but problems. A customs official found the blueprints for the pen in his suitcase and declared, "This is the design for a torpedo!" Bíró had to produce the patents to prove he wasn't dealing in armaments. But he made it out, and he never looked back. ("He left, he didn't die here.") Even after the war he would write to Imre Gellért, "I have sworn never to go back home; this is a rational, not an emotional decision. Hungary is a poisoned land, where fascism can raise its ugly head again at any moment."

The inventor and the promoter: Bíró and Janos Meyne at
Mar del Plata, a seaside resort 400 kilometres south of Buenos Aires.
© Fundación Bíró, courtesy of Mariana Bíró.

Bíró in Buenos Aires

I

When Bíró and his friend Janos Meyne got off the boat in Buenos Aires, they were taken in Lang's Packard to the plush offices of Lang & Co., down on Reconquista in the financial district, and given a bathroom six metres square in which to work. Lajos Lang was a Hungarian businessman who had recently married the vivacious divorcée Maria Pogany—an old friend of Bíró and a dental patient of Georg, who had introduced her to the pen. She was wildly enthusiastic, her husband not so sure. In fairness to Lang, who was bankrolling the project, Bíró had arrived with no actual pens. When he asked Bíró, "Where are the materials to manufacture the ballpoint pen?," Bíró pointed to his head and said, "Here." Maybe Lang's offer of space was a comment.

My arrival in Buenos Aires was simpler, and my work space more expansive. I was travelling alone, and the apartment, once a tango studio, was easily ten times bigger than Bíró's first bathroom-laboratory. The place was, as advertised, in the heart of the city, at the corner of Callao and Peron, and though I was three floors up I could still taste the exhaust and hear buses grinding, horns honking, engines roaring. So this was Buenos Aires, the city named for "good air." Good joke. I had a dining room table that seated fourteen; I felt like a lone pebble rattling around in a tin can. I went out on the balcony and watched beginner couples stepping through tango moves in the studio across the street. I had only the shred of a lead for my research: a blurry photograph off the internet that showed Bíró's daughter, Mariana, presiding over some school occasion a few years back. She had a vague smile. The smile of the often-photographed? Or the smile of dementia? Had she been trotted out as a figurehead? And was she even alive? I had tried the phone number of the school, Escuela del Sol, from Canada but it didn't connect. I tried from the apartment and this time got through, explaining slowly in English that I was a Canadian professor and wished to speak to Mariana Bíró. "No, we have no Canadian professors here," the woman said in Spanish, and hung up.

The next day I tried the Library of Congress, but it was closed, with notices plastered on its front. The block reeked of human urine, and a guy smiled and said something to me as he got into a big cardboard box and closed the flaps. It didn't look like the library would open any time soon. I walked up Avenida Florida, where every third store was a leather shop. All the men wore suits and the women stylish dresses, and everyone carried beautiful leather portfolios or purses. No academics with ballistic nylon shoulder bags. The long hike up Avenida Santa Fe took me through upscale Recoleta into even more upscale Palermo, past impossibly stylish boutiques, clothing from edgy to elegant, and shoes of lustrous leather I'd seen nowhere except Milan.

Yet there was a softness to the street, the sun diffused through light-purple jacaranda trees, and I was pleased to see fountain pens in many store windows—working pens like mine. Then I found the fabulous El Ateneo, a bookstore in a converted century-old theatre, done in cream and gold trim, with three balconies, brass rails and burgundy felt everywhere. In the English section I found Jorge Luis Borges and bought a small volume of his late short stories, *Brodie's Report*. I dipped into it, then bought a coffee, enraptured. A city of bookstores and pens and good cafés. Bíró had landed well.

Initially, Meyne did not believe in the pen—it was an excuse to get out of Paris—but he became indispensable as Bíró's promoter. A flamboyant figure (and according to Moldova something of a gigolo), he masterminded the advertising that attracted the attention of George Martin and the London Bank, which led to the Royal Air Force's buying the pen. After the first pens foundered—the "Eterpen" that was to write eternally—they needed a new name; Bíró insisted that Meyne be included, and the company became Birome—Biro and Meyne. It was Meyne who created a rumour that the peace treaty to end the Second World War would be signed with a Birome. And, always the ladies' man, it was Meyne who came up with the idea of setting up a customer complaint office staffed with attractive women to exchange defective pens or to arrange for the cleaning of clothes damaged by ink leaks. Bíró himself was something of a flirt; in his memoir he speaks of meeting the seventeen-year-old daughter of a French diplomat on the boat and teaching her how to kiss.

I found the site of Bíró's first workshop at 3040 Calle Oro, just off Avenida del Libertador; I took a picture of the brown metal door and then strolled down to the big Tres de Febrero park. I passed an armoured jeep with two guys chatting, one in a black turtleneck, the other in shirt and jeans. They looked like two young men talking about girls or cars, ordinary except that the jeep had its windows and windshield replaced with steel and everything painted matte black, including the nozzles of the four guns on top. But the park was beautiful, lush, and quiet; Bíró would have had red-earth paths, palm trees, and birdsong not far from his door.

II

Bíró did not take many breaks, for the right ink still eluded him. "I locked myself in the lab and spent days and nights trying out different substances," he writes. His brother Georg was still sending suggestions from Budapest—none of which worked. Georg lacked the inventor's sense of improvisation; it was like a jazz musician trying to collaborate with a competent but uninspired classical player. "Docile and susceptible to my mother's conservative influence, he latched onto concepts already elaborated and refused to try new ideas. He would memorize his lessons without thinking critically," Bíró remembered. When he came to Buenos Aires in 1941, Georg "started by using oil and resin and persisted in trying thousands of combinations of oil and resin. If he had not become obsessed with these two ingredients, we would have had a breakthrough sooner, and sooner or later arrived at the ink made with glycols, which is the one used today."

Georg's one-track mind was maddening, but at least he didn't come to the lab. Maria Lang, however, was always around. Bíró couldn't dismiss her because she controlled the flow of cash; and, a further complication, she was sexually alluring. As a girl, he notes, "she was very attractive but also very shallow, her conversation barely above the frivolous stories found in women's magazines." By the time they met again in Buenos Aires she had become "an exceptional, strange woman. She lived in a fantasy world that to her was as real as her daily life." She would come to the factory and interrupt him with "mysterious stories where she was always the heroine…. All of this was mixed up in a sentimentalism that was especially seductive." Where Georg had too little imagination, Maria Lang had too much, and she'd taken it upon herself to supervise the development of the ink. "One night she called and told me to stop all my work. She was sending over the ingredient that we had been searching for all along. It would make the pen work. I begged her to tell me what it could possibly be. It was Pond's cream! She had 'discovered' it while putting some on her face. She was sure it would make the ballpoint move smoothly." Bíró knew it wouldn't work, but she controlled the money, and so, he wrote, "I was forced to put some inside a few pens and test them. Needless to say, it was a disaster."

Maria Lang was both a seductive dreamer and "an able businesswoman and negotiator who mixed the commerce of goods with the commerce of souls." Bíró had already been forced into a bad bargain with the Langs—in return for their backing they didn't split the profits 50/50, as he'd anticipated, but in three equal shares. "It was useless to argue that it was illogical for Mrs. Lang's portion to be equal to mine, the inventor and developer. Neither did it seem fair that I should have to pay both Meyne and Georg." But in August 1941 Hungary had passed the third Jewish law (making marriage or sexual relations between Jews and non-Jews illegal), and Bíró wanted to get his family out. The Langs agreed to bring over his wife, Elsa, and Mariana, plus Georg, in return for half of Bíró's remaining third. He didn't hesitate: "It was the best deal of my life! In exchange for some figures on a piece of paper, I would get human lives, the most precious thing imaginable. What greater joy could I have? My happiness was obscured only by my worry that it would be too late to get them out of the Nazi hell." But once they were all safely in Buenos Aires, Georg got on his nerves: "The very next day I was back in the lab working on the ink. Georg was not a part of our staff and went to work for IMPA, setting up a factory to manufacture cellulose acetate. He earned a good living—which he naturally kept for himself—but he soon lost the job when the company was taken over." Bíró is resolutely sunny, but you can feel the edge in that "naturally" and a hint of *schadenfreude* in "he soon lost the job."

Work continued into 1942, and Bíró at least figured out a way to keep the ball from jumping out when he pressed on the piston. Bíró's suppliers in Sweden told him that no one had ever requested balls of such perfect precision or tiny diameter. They modified their machines to accommodate him, and to test the balls he used parts from a record player to create a machine that moved the pen over a strip of slowly moving paper, a version of which is still used today. Encouraged, Meyne launched spectacular double-page ads in all the major newspapers announcing the ETERPEN: the name, said Bíró, to give the impression of "writing forever," and the accompanying picture of a military tank with wings to give the impression that it could move on land and air. Their slogan proclaimed "Eterpen, two in one: pencil and pen." But as yet it was neither—the balls wrote unevenly and the ink still leaked and hardened.

III

I wanted to find that Eterpen ad. My landlord had given me the name of a translator, but she hadn't returned my call. To combat the Friday afternoon funk I walked up to old Palermo to the house of Borges's infancy. The only trace of Borges was a plaque on the wall beside the sign of the Malditto Frizz hairdressing salon—did this mean "bad talk frizz"? I wished I'd spent time studying Spanish.

Even more depressing was Che's house around the corner, next to the Homer internet café, with a logo of Homer Simpson eating a doughnut. I took a snapshot anyway and walked up the street to Plaza Serrano.

They could have crossed paths here, the three revolutionaries. On Christmas Eve in 1938, as Bíró was making arrangements to flee to Paris and begin a new life, Jorge Luis Borges was falling down stairs, contracting the head injury that in recovery would lead to a new style, to "Pierre Menard, Author of the Quixote," published in May 1939, as Bíró was making his way to Barcelona. In 1941, as Borges was publishing *Garden of the Forking Paths*, Bíró's wife Elsa, daughter Mariana, and Georg were setting sail from Bilbao. Everyone was disgruntled—Borges's book had failed to win any prizes and Victoria Ocampo had taken up the cause in her magazine *Sur*. Georg, mistaken for a spy, was detained in Barbados for two weeks. Che at this point was only twelve, but he'd learned to bicycle in the Palermo gardens and had begun his voracious reading, which included Whitman, Kipling, Faulkner, and Camus in addition to Spanish authors. By 1945 the pen was in production and Che was in university. Bíró was still living in the area. Borges, who might have come up to refresh his impressions of Palermo for his stories, could have been sitting at one table while Che and his medical school buddies talked politics and motorcycles at another, Bíró and Meyne discussing new markets for the pen at a third.

The café on the square was packed with well-heeled tourists (I was wearing no-heeled sport shoes) and stylish *porteños* ("people of the port," as Buenos Aireians are called). I wandered aimlessly, feeling out of place with the party crowd at Plaza Serrano, wondering if this trip was going to yield anything at all but good desserts and long walks. Bíró would have gotten down to work, chain-smoking his Craven A cigarettes far into the night. What was

I doing here? Yet when I got back to the apartment the translator phoned: she would be happy to help, the work sounded fascinating, in fact she knew of a famous television host who said that his father, not Bíró, had invented the ballpoint; she would come to the apartment tomorrow morning and phone Mariana Bíró from there. Things were looking up.

I went to an *asado* place on Corrientes, where I was served by middle-aged men in elegant suits who treated me as if I were a person of quality rather than a crass *norte americano* eating two hours earlier than any civilized person. The meal was simple and delicious: barbecued beef, plain potatoes, and spinach, with a glass of the house wine. The waiter poured generously—the glass was really a goblet—and about halfway through I could tell I was getting a bit sozzled, with that pleasant narrowing and sharpening of focus that wine sometimes brings. I took in the heavy linen tablecloth, the worn but solid silverware, the deep chocolate *helados* in the wafer cup. Afterward I strolled up Corrientes, and the faces on the street seemed to float toward me. For the first time since arriving in Buenos Aires I slowed down, took in the crowd instead of trying to cut through it, lounged through the dusk.

As I walked I thought about Bíró's second "eureka" moment, less a bolt of insight than a shift in perspective. It had come at a café, a roadside far removed from the sophisticated cafés of Budapest. For six years Bíró had been experimenting with pressure as the delivery mechanism for the ink, and nothing had worked. The constant worry about finances had, he writes, "created such tension as to limit my creative capabilities." He seldom took holidays, but when Meyne got an offer to buy land in northwest Argentina, Bíró decided to go with him. En route to Catamarca province they stopped to eat at a restaurant in Tafi Viejo, a small town fifteen kilometres north of San Miguel de Tucuman, which as far as I know has no statue of Bíró or the pen, but should, because this is where his decade of work on the ballpoint crystallized. He records the scene in *Una Revolucion Silenciosa*:

> Entering the patio I saw a girl of about fourteen or fifteen, sitting on a little stool in front of a stone table with a concave opening in the middle. She was beating something in the hole with a round rock.
>
> I asked the owner of the store what the girl in the patio was doing.
>
> "She's my daughter. She grinding corn to make flour."
>
> "But why doesn't she use one of these modern mills that you have on the shelf?" I was intrigued.
>
> "Why would she do that? She has plenty of time!"
>
> We ate and returned to the store to pay for our meal. We saw the same scene as before. The Indian leaning on the counter, his fingers moving slowly over the same fabric. I realized that, for them, time had stood still. I thought about our pendulum clock and realized that its march is continuous but the time that it measures is relative. From that moment I felt an immense peace and my tension disappeared as if by magic.

As at the Japan Café, this click of the mind, this little shift out of the accustomed grooves—the thing Georg never permitted himself—allowed a new concept to slide in: capillary action.

Bíró writes, "If the ink has a determined viscosity and tension and if the space between the ball and setting are filled completely, a capillary force is at work that counteracts the force of gravity, that acts against the column of ink. This was the birth of the capillary system of feeding ink." Bernie Zubrowski's 1978 *Ball-Point Pens* (a

Children's Museum Activity Book) puts it more clearly. Capillary action is the tendency for liquids to creep into narrow spaces—it's what makes candle wicks and paper towels work, and what draws ink to fountain pen nibs from the reservoir in the pen. A combination of surface tension and the adhesion between the liquid and the sides of the container will act to lift the liquid, counteracting the force of gravity. The problem for the Bíró brothers was to find a way to keep the feed constant, yet keep the ink from oozing out around the ball. Bíró fed the ink from a large tube into a narrower tube, thus creating sufficient capillary action to draw the ink down continuously, keeping the ball moist. At the same time the capillary action kept the ink from oozing out.

The Biro De Luxe: at £5, it cost the equivalent of half a week's wages in 1949 Britain. Photo by Jeff Papineau, Special Collections, University of Alberta, courtesy of the author.

Moldova has a less romantic version of the story: Bíró, preoccupied, put a pen and cartridge on the test machine without the plunger, forgot about it, and then found that the pen wrote anyway. This caused him to start thinking about capillary action. I much preferred the story of the girl grinding corn. I had come to accept that inventors were imaginative and secretive, and crafted creation myths to go along with their devices, blurring the trail of development.

When I reached the apartment I settled in with my volume of Borges stories (I like reading when I'm a bit drunk), and found, in "Juan Murana," this passage: "Your father, rest his soul, told us once that time can't be measured in days the way money is measured in pesos and centavos, because all pesos are equal, while every day, perhaps every hour, is different...." I went out onto the balcony to close the steel shutters and thought, It could be worse, having to spend a Saturday night reading Borges in Buenos Aires—the traffic going by, that once-intolerable noise that now sat on the edge of my consciousness, the girls from the tango studio sitting in the window smoking. The perfect place to read Borges's stories of knife fights in Palermo and San Telmo, places where I'd walked during the days. Maybe I shouldn't be so anxious; maybe, as for Bíró, things were unfolding as they should.

IV

When the landlord had said he could arrange a translator I instantly imagined someone young and lovely, and then stopped myself. It wouldn't be that way; it never is, and anyway I wasn't interested. I conjured up instead an earnest linguist with thick glasses and stumpy shoes. When the thoughts returned I added stocky, swarthy, and a fierce command of the subjunctive in six languages. On the morning she was to arrive I dialed in a hint of a moustache and sour body odour.

The doorbell startled me: I'd expected the intercom buzzer from the street. I opened the door and peered out to the darkened landing. A voice said, "Someone let me in below. You are Professor Bishop?" I said I was.

"I am Lucila," she said, extending a hand and stepping into the light. She looked like Audrey Hepburn. Slender, with brown hair that fell below her shoulder blades, held back in front with a hair band. No makeup, delicate features, small hands. She wore tight blue jeans and a western shirt with snaps over a turquoise tank top.

I asked her to try phoning the Escuela del Sol. She took my pen to copy the number, but it slipped from her fingers and hit the tiles, the nib skittering across the floor. "It's okay," I said, "it's cheap and they sell them everywhere here."

"No," she said, "this has happened with mine; I can fix it." She slotted the nib back on and tried a couple of letters. "See?" She smiled.

Lucila dialed the number and, holding the phone, she looked at me and fluttered her hand at her chest—she was being put through. "*Si … si …* "

After she hung up she told me, "That was Mariana. She said, 'My father was a journalist and I know it is hard to get interviews—why don't you come at nine tomorrow morning?'" We were in.

We spent the rest of the day together at the Congressional Library, looking through *La Prensa* from the 1940s. Lucila had never done this kind of work before, but she turned the big acid-saturated sheets—like giant autumn leaves—delicately. We found only a few fountain pen ads, but they provided useful context. Lucila said, "The word they used for making the pen is a word we use in Spanish for buildings: not something small, it's for something monumental. And this ad says, 'Do your children deserve a Waterman?' A pen was important."

Over lunch Lucila told me she'd done a degree in translation and before that a certificate in tourism and hotel management. She worked half days for American Airlines at the airport. "I like that because then I can use the rest of the day for doing other things, like giving tours for a tour company. Later, if you have time, I will give you a tour," she said. "Be careful of your shoulder bag. Always keep it beside your feet where you can feel it. They don't mean to kill you, but often they are drug addicts and things go wrong. My supervisor at work, her husband was killed. They were in the car close to home, it was nighttime, and another car headed them off. It was a carjacking. Four men jumped in and one had a gun; it went off and her husband was shot in the heart and died.

"Where my parents live is a middle-class suburb, and they have security guards on the corner. But I don't want to make you nervous. I have lived here many years and I have only been robbed once." Bíró never talked about the violence, but as I would soon learn from his daughter, it had always been part of the background in Buenos Aires.

Mariana Bíró at the Escuela del Sol, Buenos Aires. The proportions of her father's pens inspired the Fisher Space Pen. Photo courtesy of the author.

Mariana

I met Lucila at the Callao station for the green line at eight a.m. I had on dark-blue pants and a light-blue short-sleeved shirt and was carrying my blue cord sports jacket, even though it was too warm for it. I wondered if that would be formal enough. Should I have worn a tie? "I think it's okay," said Lucila.

We had *medialunas* and coffee at a café near the school (I didn't want to be rattled by hunger partway through the interview, though I had no idea how long it would last), then walked over and rang the buzzer at the gate. I could see kids tearing around on the other side.

"This is very unusual," said Lucila.

"What?"

"The children have no uniforms. All private schools have uniforms."

It didn't seem odd to me, and I forgot all about it as a woman with reddish-brown hair in a light denim shirt and dark-blue pants came up and put her hand on Lucila's arm. She said, smiling, "You must be the translator." The joke was that Mariana needed no translator: she'd married an American and spoke perfect English. She took us to her office and sat us at a round table that filled the small room. "I'll go around," she said, and disappeared back out the door we had entered and came in another one from the side, behind the table. It was like a magic trick. She was fluid and agile, with none of the creakiness or cautiousness I expect to have in my seventies, and agile in speech as well.

She opened with the story—doubtless told hundreds of times—of how her father came to invent the ballpoint pen: "When he was a journalist he would do an interview and his fountain pen would spot the paper.... They told him he was crazy: 'Do you want to write with a little ball?' ... 'And the ink, it's either going to be wet or it's going to be dry' ... The first pens spotted everybody's shirt ... the factory had a special office where you could go if your clothes got ink stained." At this point a boy with a newly fallen-out tooth came in, and Mariana attended to him. It turned out that he lived in the same apartment Mariana and her parents had lived in when they first arrived in Buenos Aires. I took this as a good omen—we were all connected by Bíró.

Mariana took up one of her father's pens from a rack on her desk. The remarkable thing about the Biro pen, she pointed out, is how unremarkable it is. If you saw a car from the 1940s, or a radio, or a washing machine, you'd instantly recognize the era. As for computers, they've changed more in six years than the ballpoint has changed in sixty. "It had to have a certain weight, a certain thickness, so it was comfortable to write, and they're still modern now. The Biro was the model for the Fisher Space Pen," she said.

Bíró created over thirty successful inventions, from the domestic to the industrial, including an unbreakable door hook, a heatproof tile, a watch-like device that measured blood pressure, the cigarette holder, his automatic transmission, and an electro-magnetic mail system that employed the principle later used in Japanese bullet trains. At the end of his life he was working in the garage of his home on a gas-separation process to enrich uranium for the Argentine government. Yet the only thing besides the pen that he put into production was the cigarette holder.

"The inventor's mind is so ample. My father never wanted to be an engineer because he said, 'Once you learn how something works, that's the way you think it works.' Inventors say it is usually about six years between the conception of an idea and the patent. He was not an industrialist, he was an inventor. The question for inventors is 'Can this be done?' Even if he never wants to manufacture it. My father always put his money—and he made a lot over the years—back into his inventions. If you ask an inventor what his best invention is, he will always say, 'The next one.'"

Then Mariana asked me what I'd done in Buenos Aires. This was awkward. "Well, not much actually. I've mainly been walking." I started to make my visit to the library sound bigger than it was, and then just told her the truth: that when I'm in a new city, I like to walk it. I told her how I'd walked up Florida Street in the rain the first day, discovered that Borges's house was now a parking garage, then explored San Telmo and La Boca one day, Palermo the next … that I'd discovered some good cafés, established that the Martinez cafés weren't great … had been dazzled by the lavish El Ateneo bookstore built in an old theatre …

She looked at me. This is it, I thought, the end of the interview. She's a powerhouse who at seventy-five can run a school and I'm a deadbeat who can't even expedite a simple research project.

"You know," she said, "the first time I went to New York was with my father. I was fourteen years old. He spoke Hungarian, he spoke German, he spoke French, he spoke Spanish, but he didn't speak English. The technical language he knew about, not the conversation part, so he said, 'Why don't you come with me and be my translator?' We took this plane which was a DC 3, Pan American, and it took us four days. Twin engine, and it was twenty-one passengers, and when you walked in, it wasn't level. The first night—they didn't fly at night—we stayed in Rio, the second night in Belem, the third night in Miami, and the fourth day we arrived in New York. It was quite an adventure.

"And you know, you flew low, so you could see things. We flew over the Amazon, we flew over the Cataratas del Iguazu. You were right there. It was wonderful. We stopped everywhere.

"Anyway, we got to New York and I remember we were staying at the Plaza Hotel, and it was full of people who could translate and secretaries, and he sort of forgot that I was around, and he said, 'Why don't you go around and walk. Walk.' I was fourteen. 'Walk in New York. Get to see New York.'

"And I said, 'If I get lost?'

"'It's impossible to get lost. Go down and you'll see Fifty-Nine and Five. It's all numbered, the streets. So come back to the same number.'

"So he gave me some change and some money and I started. I walked for a month. Every single day I walked all over New York. I love New York. It's a city with a lot of energy. It gives everything, doesn't ask you for anything.… So New York is mine, in a way even more than Buenos Aires."

Walking in the city. I had said precisely the right thing. Mariana had already given us coffee, but now she asked, "Do you know where the best empanadas in Buenos Aires are made? No? Here at the school." A few minutes later a child came in with a plate of empanadas. They were delicious, and as Lucila and I ate with our fingers, Mariana went back further.

"I was nine years old and I wanted to learn Spanish badly because I couldn't communicate with anyone. I took a Spanish-Hungarian dictionary and the newspaper and crossed to the park in Palermo. I used to go to a place where they had benches and read *Popeye, Flash Gordon* … there were no comics in Hungary … and I'd look every word up."

She loved helping her father. "The first rudimentary samples were really not working, and I was paid fifty cents an hour to try them out. I usually did about two hours work in a day. Then we got this metal hand that held the pen." And the ink wasn't the only problem. "I remember the first ones were plastic, and plastic had urea in it, and they always smelled like pee when it was hot." (I later acquired a vintage urea Biro—when you take the cap off it knocks your head back.)

I asked about her father's lawsuit over the use of the Biro name for Bic pens.

"My father was in bed with a cold, and Meyne came by and he says, 'You know, I was in the States and I bought this package of pens, and it says Bic by Biro. Did you ever give permission to Bic to use your name?' My father had a terrible memory.

"'Ack,' he said, 'I can't remember if I ever did. Why?'

"'Because they can't use it, your name, unless you gave them permission.'

"He said, 'No?'

"Meyne said, 'No, of course not. I'm going to talk to lawyers.'

"So Meyne talked to a lawyer who said it's true, they can't use 'Biro' without permission.

"My father said, 'Who cares?'

"Meyne said, 'No, you should care,' and he started the lawsuit. Then my father died. Meyne carried on with the suit and then he died.

"Then I had this suit going on that I knew nothing about. I certainly didn't care. Lawyers here, lawyers there, letters going, letters coming. Finally I wrote a letter to Bruno Bich and I said, 'Look, my father invented the pen, your father put a cheap pen on the market, why should we fight?'

"And he wrote back, 'You're absolutely right. Come and we'll talk.' So I went to New York and I saw Bich at the factory, and we talked. That was it. A silly fight."

Fine, I thought, but Bíró opens his memoir by saying, "I feel then that the main character of my existence is not me, only a tool used to write called the ballpoint. The Birome takes the place of Biro, and I am converted into its accessory, like the ball or the ink." He knows the pen is the thing that defines him. Surely Meyne is right, the name must count. But maybe Bíró is right. His name is out there. And every hour spent on a court case, every hour spent thinking about a court case, is a creative hour lost; anyone who's engaged in a lawsuit knows how it infects your brain.

"My father certainly never fought in his life," said Mariana. "He didn't have time for that."

But what about Andor Goy? I told her I'd been reading Moldova's book.

"Oh, Moldova. He wasn't pink, he was communist red. He seemed to think my father was gypped. I don't think he was gypped. The patent lasts seventeen years or whatever it is and then it becomes public, as it should.

"Andor Goy was a very nice man whom my father worked with. I met Mrs. Goy too when I went back to Hungary. She was very similar to my mother, and though they had never known each other, we talked about the similar experiences of living with an inventor. I met the daughters, too; they were very nice people.

"My father used his installations for his experiments with

the pen, but Mr. Goy was mainly involved with typewriters. He believed he had the rights, but he didn't. Henry Martin, the British banker, was a son-in-law of Bich and he had the rights."

Thus we dismiss years of international dispute and litigation.

Yet Bíró portrays himself as betrayed by Goy, bamboozled by the Langs, cheated by Milton Reynolds (the flim-flam artist; more on him shortly), and finally swindled by Maria Lang on her deathbed. When he was going to be out of the country for a shareholders' meeting she asked him to transfer voting rights— which effectively transferred ownership of all his stock to her. She refused to transfer the shares back, yet when she was dying she had a change of heart and told Bíró she would leave him the stock in her will. Instead she left it to her son from her first marriage, Harry Kleinlein, who professed to be anarchist and anti-capitalist, but, as Moldova says, "A vegetarian soon alters his ways when he inherits a sausage factory," and Kleinlein refused Bíró's appeal. Bíró wound up with no shares in the company where he went to work every day, the company that was producing his invention.

Lucila asked if Bíró felt more Hungarian or more Argentine. Mariana said, "Oh, he loved this country, and we did become Argentine citizens when we emigrated. But nationalism never meant much in our family. I think maybe he thought of himself as a citizen of the world. Argentina took us in, it gave us a lot—but I think we gave a lot back to Argentina.

"What compensates in this country is the humanity. They are very warm and friendly.... But we have no sense of community. 'Us' doesn't exist. It's 'I,' 'I,' 'I' ... very individualistic.... We work on that in school. But the society is that way. It's hard to transform."

"What about the uniforms?" Lucila asked.

Mariana said, "Oh, who am I to tell four hundred parents how they should dress their kids in the morning?"

This sounded casual, but it was a direct challenge to the accepted order. Later, after coffee, Mariana talked about the Dirty War, when some of her students disappeared. (Some thirty thousand Argentines were killed or "disappeared" in government repression of the left between 1976 and 1983.) "We worried because here we were, a school run democratically, with elections and without uniforms, with less discipline. I think because my husband was American they thought we had special protection from the American government, and we weren't going to tell them otherwise. We protected the students. We have many stories from that time.

"Once a student came and said to my husband, 'Mike, I think they're investigating me,' and Mike said, 'Come in, you'll be safe here,' and he kept him for a while and then thought, How am I going to get him back to his house? And so he called a moving company that we often use; they sent around a truck with a big box, and the boy got in the box and the truck took him home. That was a funny one."

They even had the son of one of the generals in the school while students were disappearing. "It was a bad time," said Mariana. "You don't realize when you're living through it. But later, looking back. The thing is you see one sign, and then another, and after a while you get used to it, and that's the worst." Her face clouded (I'd always thought that was a figure of speech, but it did darken) and she said to Lucila, "That was a difficult time." As in Budapest, I had the sense of being out of my depth.

Now Mariana works to build community and civic respons-ibility. "You can shoot your mother in front of witnesses, and you won't go to jail because it's Monday, or the judge is busy, or whatever. The kids see this, and they come to me and say, 'Why does this happen?,' and I have no real answer for them. I tell them that it's the law now, and they have to obey it, and if they think it's a bad law they should become lawyers and change it. But it's very hard."

She stood up and invited us to see a classroom. We passed a poster on the wall showing a young man beside a wheeled contraption with a propeller. The slogan in English read, "Kids are inventors."

"My father said, 'Try not to teach children. Don't spoil them by teaching them…. We tell our teachers, 'Don't teach, allow the children to learn' … What's instinctive in a teacher is she wants to tell you what is right and what is wrong. They have to learn the names of the rivers in Africa. Okay. But it's the man who lives beside the river that's important."

She showed us some notebooks in which the kids were rewriting the story of Little Red Riding Hood. "Wolves are not always bad, eh? And what else could she have done with the grandmother?"

I asked Mariana if I could take her picture; she laughed and said yes, as long as I made her look "thin, tall, and young." I assured her that I'd put the camera on those settings.

When I transcribed the tape, I was amazed at how little it conveyed of the excitement of actually meeting Mariana, the immediate ease we all felt with each other, the way Lucila and I, though we'd known each other less than twenty-four hours when we went in, felt like a team. Mariana had said, "My father was interested in the transmission and conservation of energy, and the sense of what happens between two people. Why they sympathize the way they do…. You walk into a room and there's tension; you can sense it. It's energy in some form…." What the tape did catch was the background the ear weeded out: the joyous, raucous noise of the school, loud and constant. There was no better testament to Mariana's success than all that racket.

Mariana lent me a DVD of a TV program in which she's interviewed about her father. In it she says he was in a lot of pain at the end, and that Bíró had said that when a person is in pain they can't think, and when they can't think they're not alive. "You may do with this information what you want," he'd said to the doctors. Then she says her father died of ink poisoning. Ink poisoning! This was a research coup (scholars live for these moments: the chance to push back the boundaries of knowledge if only by half a millimetre). It was also like opera or melodrama. The inventor, after all those years trying to find the right formula, is poisoned by his creation. But by what exactly? The fumes, or maybe chemicals absorbed through his fingers?

FOUR DAYS LATER I went to lunch at Mariana's place up in Colegiales, on the north side of Palermo, a beautiful white townhouse with a big palm tree in the garden. "It's the house I grew up in," she said, "and after my parents died I was going to sell it, but then my daughter wanted my apartment and said, 'I'll move you back in.' So here we are." By the front door is a white marble plaque, installed by the Argentine government on the hundredth anniversary of her father's birthday. His name looked very Spanish to me:

LADISLAO JOSÉ BIRÓ
(29-9-1899 – 24-10-1985)
INVENTOR DEL BOLIGRAFO
VIVÓ IN ESTA CASA DESDE 1948 HASTA 1985

In the house the first impression is one of light. Light pours down the central stairwell from the skylight three levels above. In the living room are beige couches against white walls. When you pass into the sunroom off the kitchen, floor-to-ceiling windows take you into the backyard, an oasis with a big palm tree on one side and a tall dark cypress on the other, with lush bushes across the back wall. A white chaise lounge and two white folding chairs flanked by drink tables sat in the shade. This was no inventor's bohemia. Bíró wasn't mixing ink samples in a garret.

Mariana had invited Oracio, the former Argentine ambassador to Budapest, and Marta, his wife. Marta wore a long green skirt and wool top with a gold pendant, very simple but elegant. Oracio wore a green cotton shirt and tan pants that perfectly complemented his wife. The two had easy smiles and sat knee to knee on the couch. They had great affection for Mariana. "She works too hard," said Marta when Mariana was out of the room. "She should slow down."

We began with an appetizer of grapefruit sections, blue cheese, and nuts. Oracio said, "Mariana has a genius for putting together unusual salads." I waited until the main course came and then mentioned the DVD, and her father's death. "Oh yes, that DVD. I'm glad you liked it, but you know, I lent it to some American station, and when they finished they said, 'Is there anything we can do for you?' 'Yes,' I said, 'it would be nice if you could put English subtitles on it.' So they did, but somebody misunderstood the Hungarian, or I don't know what, and said my father died of ink poisoning. Ink poisoning!" She laughed. "We don't know what he died of," she said. "It wasn't lung cancer, which was surprising because he smoked a hundred cigarettes a day."

I felt relief. Now that I thought about it, for Bíró to have died of ink poisoning would have been corny, not mythic.

Now Mariana was talking about the cigarettes. "I grew up with smoke, and ashtrays all over. He'd start one cigarette and put it down and forget where he put it and light another. And when I'd bring my report card home, I'd bring it to him to sign, and I'd put it on the table, open, to show him, and he'd say, 'What is this?' and I'd say, 'It's my report card, see? You have to sign it.' He wouldn't look at it. He'd close it and say, 'Are you happy at school?' I'd say, 'Yes, I'm happy, but look how I've improved in these subjects,' pointing, and he'd say, 'I don't care, I just care if you're happy.' And all this while the cigarette ash was getting longer and dropping on the report card. Every one of my report cards has burn marks in it.

"Everyone tried to get him to smoke less. But when they said 'You smoke too much' he would say, 'Yes, but I drink a lot of coffee.' And when they said 'You drink too much coffee'—because he did—he'd say, 'Yes, but I smoke.' At age seventy he cut down from one hundred cigarettes a day to sixty. And lived another fifteen years."

It was the cigarettes that, indirectly, led to his visiting Hungary once more. Elsa and Mariana badgered him constantly about his smoking, so he invented a double-filter cigarette holder that, he claimed, filtered out seventy-four to seventy-eight percent of the harmful elements in the smoke. He sent a sample to General Juan Perón, who invited him to Casa Rosada (the Pink House, the official executive mansion of Argentina), where Perón persuaded Bíró to undertake a trade mission to Hungary. So off he went, back to the country he hadn't seen since before the war. But much had changed in Budapest, and Bíró no longer felt at home.

Bíró always kept a low profile, but after he got back to Buenos Aires a journalist tracked him down and asked him if he'd been honoured in Hungary. Bíró said no, but that he didn't define success in those terms. What then, the journalist asked, was his measure of success? Bíró said, "The fact that you are holding a ballpoint pen in your hand as we speak. It means that millions of people can get hold of a pen cheaply and easily. Just as much a boon for democracy as the transistor radio or the quartz clock. The bastions of privilege based on money have fallen."

After lunch we moved into the living room and sat around the low coffee table. Mariana brought out a pile of papers to show me: ads and clippings as well as three black-and-white photographs of the first workshop.

There were also some nudes in the portfolio, and I wanted to ask if they were her mother, but of course the question should be irrelevant to art. He had a beautiful secretary, Mariana said. "Women were attracted to my father, but he always made it clear he had a wife and daughter.

"My poor mother, they got along extremely well, you know. He used to phone at twelve o'clock and say, 'Well, I'm coming home in half an hour and I'm bringing six people for lunch.' We'd always work it out.

"One time my mother had an operation on her eyes, and when they took the bandages off at home she said, 'Lazlo, I see people in white coats.'

"He said, 'Yes, dear.'

"'So they are real?'

"'Oh yes.'

"'Who are they?'

"'They're chemists. They are helping me with this experiment that I'm doing and it needs twenty-four-hour care.'

"'Ah, okay, where are you doing this experiment?' she said, thinking he was doing it in his lab downstairs.

"He said, 'You know, this experiment has to be run in the bathroom.' That was the bathroom off the main bedroom upstairs.

"And then she said, 'How long are they staying?'

"'Five days at most.'

"My mother said, 'That's fine.'

"They stayed two years.

"And of course that meant lunch and dinner. Years later a lady from the atomic commission came up to me and said, 'I'm the chemist who worked with your father in your home. You know, your father was a very nice person, but what I liked best was the Hungarian food that your mother cooked.' Basically she was his anchor. He didn't know how to use a hammer. He never nailed a nail in his life. My mother fixed all the irons in the home."

ANDOR GOY CONTENDED THAT Bíró had invented nothing new, except the ink, and he notes bitterly that Bíró refused to send him the formula—even in his memoir Bíró blurs the issue, revealing nothing of the chemistry or when he perfected it—but I now saw the pen differently, the creation of a gentle, obsessed, flirtatious dreamer who wanted kids to dream.

The pen would be taken up by bomber crews and become the mark of the Cold War, but it was Bíró's stroke against fascism, a pen that everyone could own. And although the Hungarians discarded him, the Argentines revered him: they put his face on a postage stamp, created an exhibit at the Argentine Society of Engineers, and when President Néstor Kirchner took office he waved away the ceremonial pen and signed his inauguration papers with a ballpoint. Bíró's birthday, September 29, is Argentina's Inventors' Day. "The Birome takes the place of Biro," Bíró had written. It was true that Biro, mispronounced Bye-ro, was the pen worldwide, yet in Argentina, unlike in Hungary, Bíró the man had become part of the culture.

One of the last things Mariana said to me was, "I remember my father, and we had very long conversations, wonderful talks about all kinds of things. And my mother would interrupt, after two hours or something … she would listen and she would say, 'You

*First sold to RAF bomber crews, after the war Biro pens were
marketed to the whole family. Photo by Jeff Papineau, Special Collections,
University of Alberta, courtesy of the author.*

know, you have these fantasies, and what about reality?' 'Reality is
always there,' he would say. 'The last thing you want to do is reality.
It's always there.… But let us live!'"

For Bíró to live was to dream, not idly but with an engineer's
passion for working out something new. Self-promotion was the
last thing on his mind. I was about to meet someone for whom it
was the only thing.

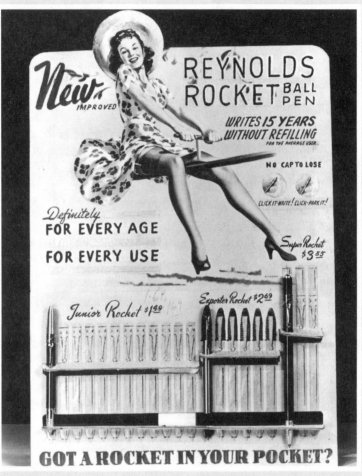

"Got a Rocket in Your Pocket?" 1947 Reynolds ad. The Reynolds campaign also featured girls in bathing suits writing underwater. Disgruntled customers complained that that was the only place the pens would write.

The Flight
of the Reynolds Rocket

I

L azlo Bíró was a dreamer, Milton Reynolds was a schemer. Young Milt started a used-tire business, and by the time he was twenty-six he was a millionaire. By thirty he was broke. He would go bankrupt three times, and each time recoup his fortune. Born in 1892 (seven years before Bíró) in Albert Lea, Minnesota, Milton Reynolds was a consummate promoter who cared nothing for craft. His father, Simon Reinsberg, had emigrated from Prussia where, like Bíró's friend Meyne, he served in a Hussarian cavalry regiment; in Minnesota he sold threshing machines. But his son whirled like a thresher through one product after another, promoting it, taking the profit, and moving on. He had a huge success with the Printasign—a device like a giant typewriter that turned out signs for department stores. Later he imported cheap cigarette lighters from Mexico. They looked like Zippos, but leaked, caught fire, and occasionally exploded. (With Reynolds, "getting burned" wasn't just a metaphor—rumour had it that one man died when a lighter blew up in his breast pocket.)

In June 1945, having acquired one of the new Biro ballpoints from a friend in the Pentagon, Reynolds "felt the heat of a strange fever" (his words): he had to have that pen. He flew to Argentina and contacted Bíró. In his unpublished autobiography, "My Pen in Hand," as told to Arnold Drake, you can hear Reynolds's voice, his flair for the dramatic:

> In the office of his Argentine company, Eterpen, the smiling Biro told me [how] he gained the idea of inserting a steel ball in the pen's socket instead of the conventional point. Liquid ink proved impractical, leaking through the minute space between ball and socket no matter how tightly they were fitted.…
>
> Biro then explained that he had gained a patent on certain elements of the pen in Paris in 1939. That phrase, certain elements, stuck in my head for some reason.… Now what was that all about? Did he have a patent or didn't he?

Bíró told him that the U.S. rights had been sold to Eberhard Faber, but Reynolds bought four pens in the airport and dashed to his plane, obsessed with the marketing opportunities. The war was winding down; the armistice would be "the most joyous occasion of the century. And I would share in that joy: … Christmas 1945 would bring the biggest buying spree in history."

Bíró's status as the inventor of the pen was for Reynolds a mere legal technicality; he instructed his attorney to check it out, and he discovered that the actual ballpoint process had been patented back in 1888 by a Massachusetts tanner, John Loud, for a roller ball that would apply ink to leather hides. So that aspect of the pen was in the public domain. However, what made the pen work was Bíró's capillary-action feeder system and his ink. Reynolds would have to come up with his own system—and he proposed to do in ninety days what had taken Bíró nine years.

Paul Fisher, a young engineer who would go on to invent the Fisher Space Pen, remembers the time: "I worked for Milton Reynolds in August, 1945. Mr. Reynolds introduced the ball pen to the world. He gave me one of the pens, and I worked with it for two days and then I reported back to him that the pens were no good at all, the basic principle wasn't any good…. It leaked out the front end, and leaked out the back end. It was really a piece of junk." Even the pens that didn't leak were no good, said Fisher, because "the ink had acid in it. A year later you could wet your thumb and transfer a signature from one sheet of paper to another. It never dried."

Reynolds didn't care. Having brought a million-dollar lawsuit against a rival and being hit with a million-dollar countersuit in return, he brought the leaky pens to market.

II

Reynolds had taken me from sunny Buenos Aires to soggy Seattle, where the rain turns Gore-Tex to tissue paper. I checked into the Silver Cloud Inn and walked along the bike trail to the University of Washington, cold rain sliding down my neck. Fit cyclists flitted by, their jaunty posture an affront to my cringing slouch. "We're practically underwater!" I wanted to shout at them. I shook myself off at the library and found the archives downstairs. The curator was expecting me. What I wasn't expecting was the five trucks (the wheeled carts libraries use) of Milton Reynolds material. Had he saved everything about himself? Almost. I would find two unpublished biographies, piles of news clippings, newsreels, home movies, correspondence, and a PhD dissertation about him. Nothing derogatory. Milt was a master at moulding his image.

Always more interested in promotion than perfection, Reynolds discovered while writing on a damp bar napkin that the pens would write through water. So he coined the slogan "It Writes Under Water!" and had pictures taken of pretty girls in bathing suits writing in swimming pools. As *Time* magazine reported ("Tempest in an Inkpot," November 12, 1945), "Thousands of people all but trampled one another last week to spend $12.50 each for a new fountain pen." With a characteristic smirk, *Time* notes that "Gimbels modestly hailed it as the 'fantastic, atomic era, miraculous pen.' It had a tiny ball bearing instead of a point, was guaranteed to need refilling only once every two years, would write under water (handy for mermaids)." Aquatic or not, the numbers are astonishing: "the store had placed an order for 50,000 pens (retail value: $625,000). At week's end, 30,000 pens (including 12,000 mail orders) had been sold."

Time also mentioned the lawsuits—"Thurman Wesley Arnold, hiring out his trust-busting talents to Reynolds, had filed suit in Wilmington's Federal court for $1,000,000 (treble damages) against Eversharp Inc. and Eberhard Faber Corp. on a familiar Arnold charge: violation of antitrust laws.... Back came a counter-claim, also for $1,000,000"—and the Mexican cigarette lighters: Reynolds "is apparently a 'stop-&-go guy,' a man who drops the item when it goes sour and turns to another." And they did go sour. The pens leaked so badly that Reynolds had a special contract with a dry cleaner in Chicago who'd learned how to take out the dye. *Time* reported that "ink oozed all over hands and paper. The ink in some pens even fermented, and blew the balls right out of the pens…. One disgruntled buyer said, 'The only way the darn thing will write is under water.'"

But buyers kept on coming. "Said one bemused pen man: 'They're like horse players. They figure they can beat the odds—and get one that works.'" Reynolds, the master of spin, stated in the company's financial report, "Frankly, one of the problems was the occasional development of a small air bubble in the barrel of the pen which sometimes prevented the free flow of ink to the ballpoint. Literally thousands of perfect pens were, therefore, returned to the retailer as being defective." The pens were perfect; they just didn't work.

In many cases the ball bearing clogged with bits of the ink's pigment and jammed; some failed to write unless held almost vertically; others "deposited a spoor of ink droppings along the paper"; still others oozed ink onto the fingers (many of the pens were improperly assembled). Consumers' Research reported that the ink "was a very fugitive dye, so fugitive indeed that it would seem that the greatest usefulness of the pen might be for persons who have reasons for wishing their writing to fade out rather quickly." Reynolds was outraged by this charge—ninety percent of dry cleaners found it impossible to remove the ink.

Those pens that didn't work failed spectacularly, and customers demanded satisfaction:

Dear Mr. Reynolds:

I am writing this in pencil, because when I last wrote with my pen the bearing suddenly popped out and flew up into my eye, and, worse, the whole two years' supply of ink followed. I'm not sure I <u>want</u> a new pen, but I do expect recompense for the enclosed (a) optometrist's bill for $10.00, (b) dry cleaner's bill for $3.00 and (c) rug-cleaner's bill for $20.00.

Some still hoped:

> *Gentlemen:*
>
> *I have had three of your pens and with <u>each</u> <u>one</u>, sooner or later when it's in my inside jacket pocket the ink busts with a loud squishy plop right out the <u>top</u> of the pen barrel, like a bomb!!! (Maybe my body heat makes the ink expand??? I sweat a lot.) Ruining three jackets, three shirts!!! I got left only one suit I got on now, and for my 46th birthday my brother in Omaha sent me another god-damn ball-point pen.... Hurry because I'm absent-minded and maybe can't keep the new pen out of my last pocket.*
>
> *Very truly ...*

Some just raged:

> *Reynolds:*
> *I am returning your pen.*
> *Stick it up your ass.*
> <u>*Side-ways*</u>*.*

What to do? Hire a team of engineers? No, hire a team of promoters. Reynolds brought out a series of snappy radio ads featuring Bob Elson, America's best-known sportscaster (so famous that even while he was serving in the navy during the war, Franklin D. Roosevelt brought him home to call the 1943 World Series). Then Reynolds commissioned a catchy theme song to go along with the ads. "Put a Rocket in Your Pocket" was sung by Jack Owens, the "Cruising Crooner," the man who "looked like Cary and sang like Bing"—though apparently Bing Crosby objected to the way Jack Owens would carry his microphone into the audience, mussing the hair of his female fans, kissing them, and even occasionally sitting on their laps. Reynolds had movies of girls in bathing suits. He also had print ads with an enraptured girl, skirt flying up above her stockings, riding a teeter-totter that looks like a pen above a phalanx of new Rockets—Junior, Senior, and Super—and the unambiguous slogan, "Got a Rocket in Your Pocket?" If you were a player, you had a ballpoint pen. But even these efforts couldn't hold back the tide of leaking ink. Reynolds's pens dropped from $12.50 to $3.85 to 94¢ to 19¢.

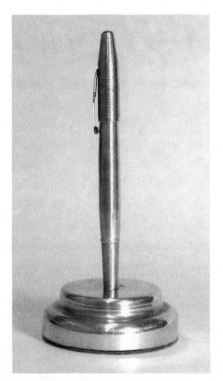

Reynolds Rocket pen: in silver steel they looked like a rocket on a launching pad. Photo by Jeff Papineau, Special Collections, University of Alberta, courtesy of the author.

What to do? Build a better pen? No, buy a bomber. Hire the famous China "Hump" flyer William P. Odom as pilot and fly around the world to beat Howard Hughes's record, handing out pens at every stop. Work in a promotional shot for Wheaties, Breakfast Cereal of Champions, and have the flight covered by Movietone News (which used to run in movie theatres before the main feature). In 1933 Wiley Post had flown solo round the world in 186 hours; in 1937 Amelia Earhart had crashed mid-Pacific trying to beat the record; in 1938 Howard Hughes had set a record of 91 hours and 14 minutes. No one had attempted it since the war—this was the record to beat. Reynolds, now fifty-four and portly, lost thirty pounds so that he could fit into the cockpit.

Acerbic but fascinated, *Time* magazine noted, "Working far better than Reynolds' pens, the Reynolds Bombshell took off from LaGuardia Field, stopped at Gander, then crossed the Atlantic in record time." On April 15, 1947, the *Lethbridge Herald* and the *Toronto Star* reported how the plane roared into Edmonton (the one Canadian stop, since Gander in Newfoundland wasn't yet part of the nation), where a thousand spectators crowded the field. The plane took on fuel while the crew took on turkey sandwiches, and sixty-nine minutes later it roared off again. The excitement is impossible to duplicate today—a Winnipeg paper found it newsworthy simply that the Bombshell might fly over their city after leaving Edmonton.

But the *Lethbridge Herald* also reported on the Ontario High Court's censuring Reynolds for failure to appear on a charge of breach of contract for a Toronto distributor. In Manitoba the law courts had stopped using the pens, and at city hall they were no longer allowed for registrations of vital statistics. In New Jersey the Treasury Department warned employees in all state offices not to use the pen. Reynolds called a press conference and signed a $100,000 cheque in a tray of water, promising to make it out to charity if after one year his signature had faded. (He knew it wouldn't because the cheque was kept in a bank vault, never exposed to sunlight.) Yet confidence was fading in both the pen and the promotion. In Reynolds's home state, the *Minneapolis Morning Tribune* ran a long feature in the form of a dialogue between an American and an uncomprehending Chinese reader from Shanghai:

> *Dear American friend, I am intrigued and yet deeply puzzled by the unbelievable haste shown by your Mr. Milton Reynolds, who made himself highly uncomfortable by flying around the world in 78 hours, 55 minutes. Why did he do this? ... Did he learn anything about the tensions in India, or in my own China? I fear not. And we learned little from him except that the plane roars and some of us have a fountain pen that writes, though I fear I cannot understand why, under water.*

Reynolds had beaten Howard Hughes, but still the pen business spiralled down.

What to do? Build a better pen? No, buy a bigger bomber. Fly across China, find the mountain higher than Everest, and name it after yourself. (Pilots flying over "The Hump" from Assam, India, to Kunming, China, had sighted a massive peak.) If Reynolds had succeeded, the Mont Reynolds might today equal the Mont Blanc in status, but the expedition and the mountain itself were, like Reynolds's pens, more hype than substance.

The flight was to be a genuine scientific expedition. Coordinated by *Life* magazine founder Henry Luce (who was a friend of Chinese president Chiang Kai-shek), the team included Bradford Washburn, an eminent mountaineer and director of Boston's

Museum of Science, as well as other scientists. When Reynolds and the expedition arrived in Shanghai, Washburn was startled to find that Reynolds seemed to care nothing for the expedition, only for publicizing his pen. He created a traffic jam in downtown Shanghai distributing free pens, and later held a press conference announcing he would soon establish a branch of his ballpoint pen factory in China. Washburn worried in his diary that there would be "plenty of trouble ahead," and was shocked when he finally saw the plane: "The camera hatch is in the side of the nose with NO HEAT at all…. No throat microphone. No electric flying suits … —BUT 750 POUNDS of pens which M.R. is already distributing freely to everyone (not the poor children!)."

Washburn called a meeting with Reynolds and told him that unless he started to treat the scientists seriously, rather than as "ten-pins," he'd have a mutiny on his hands. Reynolds promised to change his attitude. But things got worse. The pilot turned too sharply while taxiing, dropped a wheel off the runway, and the plane came to rest almost on one wing tip. Whereupon Reynolds announced that the whole expedition was cancelled. Two days later Reynolds's plane disappeared from Shanghai airport at six in the morning without an exit permit, and returned that evening after six. Reynolds said they were flying to the United States via Calcutta, but after a few hours' flying he suddenly discovered that his visa for Calcutta had expired, so they turned around. Washburn wrote in his diary, "a really fishy story." I was becoming suspicious of Reynolds myself, and began concocting my own conspiracy theory: Reynolds wasn't just an irresponsible goofball, he was a spy.

The Chinese impounded the plane. Reynolds told the airport manager that he had to top up the fuel to protect the tanks from corrosion. The next day he said he needed to start the engines to test whether they were still working well, and to collect a batch of pens from the plane. Once on the plane Odom gunned the engines and they took off, dangerously downwind, skimming the masts of the fishing boats in the Yangtze as the Chinese guards levelled machine-gun fire at their tail section. Ignoring radio messages ordering them to return and flying low to avoid the fighter squadron chasing them, they made it to Tokyo, where a jubilant Reynolds said, "Now we're back in God's country—at least Americans run it."

III

I advanced my conspiracy theory to the archivist, who clearly thought I'd been in these boxes too long. Then in Box 32 I discovered a copy of *Master Detective* from May 1950 with a picture of Odom on the front and the title story, "Airman Bill Odom was 'Marked for Murder.'" The story began with the promise that "Here, for the first time, is the story behind the Odom-Reynolds Expedition to China." There's an editor's note at the top of the article:

> … after several weeks of careful research, examination of official documents, and conversations with Milton Reynolds…. There are still some phases of it that Reynolds will not discuss. However, in the light of President Truman's announcement that the Russians are known to have exploded an atom bomb, it seems fairly certain that the real mission of this expedition was to seek this information, that the exploration was secondary. Whether the President's information that the Russians have the bomb actually came

from Reynolds has not been disclosed and probably won't
be. Reynolds himself has refused to comment on this, and
the reader must draw his own conclusion.

It's obvious what conclusions we're to draw—that Reynolds was
an important player in the Cold War and the source of President
Truman's confirmation that the Russians had the bomb. Inside are
photographs of Mount Amne Machin, of Reynolds and the crew in
front of the plane, and of President Truman giving the expedition
an official send-off—a photo that Washburn had said was faked.
There's also a full-page illustration of the highly unlikely episode in
which Reynolds confounds the submachine gun–wielding guards
by tossing a bag of pens into the crowd.

Robert Rosenberg, who wrote the dissertation on Reynolds,
was himself a small businessman, not an academic, and declares
in this afterword, "Businesses are operated for profit and in those
instances where maximization of profit and ethics are at variance,
the profit motive must prevail." Yet in the days he spent inter-
viewing Reynolds in his Mexico City mansion (with its turquoise-
blue carpeting, huge tiger skin, Chinese lacquered screens, concert
grand piano, two-storey glass-walled indoor bird cage, and bouquets
of cut flowers that perfumed the rooms), he found it "difficult to
see through his images to the hard realities beyond…. He has the
born huckster's compulsion to create everything a little bigger than
life. In minutes of personal contact he can cast a spell that leaves the
listener grasping for the facts."

Reynolds scams, skims, and moves on. Yet we're attracted to
him. He lives by his wits, falls and rises, never looks back. Rosenberg
points out that Reynolds never claimed more than that his pen was
"marketable," not that it would actually work, although "the one
claim that he made with greatest pride and the least justification is

that he was the inventor of the ball-point pen." He actually accomplished little, Rosenberg says: Eversharp would have brought their pen out shortly in any case. Reynolds was constantly mythologizing his own life, a figure like that other mythologizer from the Midwest, the Wizard of Oz.

Reynolds too was the victim of a hoax. Galen Rowell climbed Amne Machin in 1981, and when he got back to California his mother gave him a letter from a neighbour who'd been a pilot in the war and had read about his climb. He tells of flying over the area in 1944, calling out the names of the peaks as they passed:

> An Air Force captain across the aisle said, "Major, you seem to know a lot about the Himalaya. Have you heard of the peak higher than Everest?"
>
> "Oh yes … Amne Machin is one of the greatest geographical discoveries of the century!"
>
> "Well it's all a fake. We made it up. Those British correspondents kept pestering us for exciting stories to cable home, so we told them of a DC-3 that got blown off course in a terrible storm and discovered a mountain over 30,000 feet…. It was a great practical joke, and serious reports of it were published all over the world."

THE PUBLIC, initially enraptured with the idea of the ballpoint, lost confidence in the pen. Fran Seech (another Hungarian) out in California developed an ink that didn't smear or stain, and with businessman Patrick Frawley Jr. they created the Paper Mate in 1950. But they knew they faced a stiff marketing challenge, so when they were ready to crack the New York market they launched "Operation Normandy": twenty-two salesmen who visited ten thousand stores in six weeks. They would leap at the manager, scribble all over his shirt, then promise to buy him a better one if the ink didn't wash out. It always did. Better still, the pens didn't leak. Paper Mate had consolidated their beachhead.

If the fountain pen was associated with bourgeois prestige, the ballpoint was associated with war. The customs official who stopped Bíró thought the plans looked like a torpedo; Meyne advertised it in association with a flying tank; Martin bought it for bomber crews. But after Reynolds it was the child of the space race, the instrument of the Cold War. Reynolds called his pen the Rocket, a smaller one the Rockette, and they came with "rocket-launcher" pen holders. After his around-the-world flight he introduced the Reynolds Bombshell. He advertised his ballpoint as the "atomic pen," and in 1960 he even floated the idea for a pen that, instead of ink, used Polonium 208, which, he said, "will discharge alpha particles for a distance of two inches, more or less, in a straight line [and] these particles will leave a brown mark or line on the writing surface that they are guided to." Polonium is used in laser printers to propel the ink droplets and in anti-static photography brushes, but also for the triggers for atomic weapons. And, most notoriously, it was used to poison Russian spy Alexander Litvinenko in 2006. Fortunately Reynolds never put his atomic pen into production. If it had leaked radiation the way his pens leaked ink, he would have devastated thousands. As it was, he'd only poisoned the market for ballpoints.

Now I wanted to find the man whose pens would take over the world.

Straight pen, nibs, and Vere Foster copy book. Profits from the nineteenth-century writing books were used to help Irish girls immigrate to the United States. Photo courtesy of the author.

Point Man: Marcel Bich

I

Marcel Bich initially dismissed the ballpoint as a piece of junk, "a fad, a passing infatuation." He was certain that "when customers had enough of blots on their papers, when they had had enough of fingers stained by ink, they would return to the good old straight pen." This was in 1948, after the Reynolds Rocket debacle. At the end of the war, age thirty, not sure what he wanted to do, Marcel Bich had formed the Société PPA (Porte-plume, Porte-mines et Accessoires) to make straight pens (whose nibs were dipped into inkwells) and mechanical pencils. Not exciting, but they worked.

Milt Reynolds had produced nothing concrete. However, he had produced desire. He'd made pens sexy, and Marcel Bich was smart enough to see that the techno-lust generated by the ballpoint would not die down. The ballpoint had the potential to become the indispensable tool of modern times, but first it had to be made more efficient. Bich was determined to find what neither Bíró, nor

Eversharp, nor Reynolds was able to discover: the ideal formula for ink, and the perfect fit between the ball and the tube reservoir.

Unlike Reynolds, Bich was a perfectionist. He worked for two years on the ink and the machining. Finally he had the fit, and brought forth the Bic Cristal. Hexagonal like the pencil because it was better to hold in the hand, and clear so that you could follow the level of the ink, "The instrument is beautiful because it is functional," said Bich.

His wife, Laurence Bich, captures the excitement of the new device in her biography, *Le baron Bich: Un homme de pointe*: "In the past an individual might sign his or her name once or twice in a lifetime, in the register of the parish on the occasion of a birth or a marriage," but now the man of the twentieth century was "making notes on the fly, on a building site, in an office, on a workbench, on a doorstep or in the street. He had to continuously transcribe ideas, sketch the outline of a plan, respond to a questionnaire, write a cheque, sign an order book, initial a receipt...." The "habits of contemporary life," Laurence says, mean that almost everyone writes, "and not at a table or a desk." Like the phone that was once tethered to the wall, the pen had been tethered to the desk, and in this speed-obsessed, post-war age, efficiency was all. You would lose the deal if you had to run back and fill up your fountain pen.

Efficiency, divorced from sentimentality, even morality. Laurence Bich tells how "Eversharp, under licence to Martin, acquired this rare object, a stunning entry to the atomic age.... Need I say that the report of the nuclear explosion on Hiroshima was drafted straightaway in the B-29 bomber with one of the first ballpoint pens?" If the fountain pen looked like a cigar, the ballpoint looked like a missile. It was the child of the space race, the instrument of the Cold War. In fact, ten days before Reynolds's pen had gone on sale at Gimbels department store in 1945, the *New York Tribune* had featured an article by George Orwell, "You and the Atomic Bomb," in which he coined the term "cold war." Four years later, in Orwell's 1949 novel *1984*, pens and the Cold War come together. When Winston begins to keep his secret diary, he does so with a straight pen: "an archaic instrument, seldom used even for signatures, and he had procured one, furtively and with some difficulty, simply because of a feeling that the beautiful creamy paper deserved to be written on with a real nib instead of being scratched with an ink-pencil." He is unused to writing because the customary practice is to dictate into the "speak-write." Winston "dipped the pen into the ink and then faltered for just a second. A tremor had gone through his bowels. To mark the paper was the decisive act." He writes the date, "April 4, 1984"; later he will write "DOWN WITH BIG BROTHER" in full caps, and worry that the ink will mark him as traitorous:

> Two fingers of his right hand were inkstained. It was
> exactly the kind of detail that might betray you. He went
> to the bathroom and carefully scrubbed the ink away with
> the gritty dark-brown soap which rasped your skin like
> sandpaper and was therefore well adapted for this purpose.

On the other hand, Syme at the Ministry and O'Brien, Winston's torturer, both use "ink-pencils": the ballpoint pens that had just come into military use when Orwell was writing his novel. They were emblems of impersonal power, objects that were uniform, anonymous, and like the subjects in *1984*, ultimately disposable. Ink is the mark of guilt; the ballpoint kept you from getting your fingers dirty.

II

I came to Paris as I'd come to Buenos Aires—to find the site of another original pen factory, Bich's this time, and with the hope of discovering, as I had with Mariana Bíró, some connection with the past. So far the Bic company hadn't returned my calls. Hsing told me not to worry, something would turn up. We were staying in a little hotel on Isle St. Louis, where the shops were fashionable and the baguettes fabulous, but our search took us north along the Avenue de Clichy with its cheap shoe stores, cheap suit stores, and no stores selling art. The people were working Parisians, and Hsing noted lots of models—skinny young women in tight jeans or short skirts and boots, with big bags slung over their shoulders.

We stopped for coffee at the Brasserie Wepler on the Place de Clichy, a noisy, bustling intersection. We looked out at an art deco subway sign in the middle of the intersection, the one Picasso and Mondrian would have looked at as they sipped their aperitifs. Henry Miller would also have passed the café as he worked on *Tropic of Cancer*, walking home to his apartment just around the corner from where, ten years later, Marcel Bich would set up his first pen workshop. The road drifts downhill to Porte de Clichy. Gridlock, honking horns. Once you pass under the ring road and cross the Boulevard Victor Hugo you're into working-class Clichy. No sushi restaurants, no models here. Some tough guys hanging around the cemetery. Henry Miller's apartment is gone, the whole block replaced by new buildings, and when we finally found the Impasse des Cailloux the street had been boarded up at each end and the buildings razed on both sides in preparation for a condo development. So much for my pilgrimage to the site of the pen factory. Yet it gave a taste of the grit the neighbourhood must have had, and it's still a place of small businesses. A hole-in-the-wall auto repair place sported a battered Alfa Romeo sign and declared bravely, *Vente Tous Marques*. Bich's little workshop would still fit in. The place was a reminder of how the Bic, which now seems as inevitable as Kleenex, almost didn't make it.

Acceptance wasn't automatic: customers were wary and stationers didn't want the pen because it was too cheap, their markup too little. Bich had a great invention that nobody would buy. So, like Hannibal, he mounted a series of campaigns (Laurence Bich details his strategy in metaphors of war). Pierre Guichenné, president of the French Agency of Propaganda, advised Bich to use the family name but take off the *h*, to create a label that, unlike "Porteplume, Porte-mines et Accessoires," was sleek, identifiable, and pronounceable in all languages—"beek" both with and without the *h* in French. At the end of 1951, one year after the first launch in Lille, Bich had sold 21 million Bic pens and bought his first Rolls-Royce. In 1952 for the Tour de France he commissioned a truck with a giant Bic pen on the back to follow the riders. The Tour was the premier sporting event in Europe, and the message obvious: the ballpoint was hard, manly, inexhaustible. Bich opened factories in Milan and Barcelona, and a distribution centre in Switzerland; he attacked markets in Germany, Austria, Holland; he moved into Brazil to open up South America. By 1956 Bic was producing 372,000 pens per day, 82 million a year.

In 1958 Bich shifted his attention to markets in underdeveloped countries. He flew a twin-engine Piper Apache over northern Africa to check out the landscape, then sent in a fifteen-ton truck that could travel over rough terrain. It had a built-in pharmacy, and even cinema facilities. It also brought books, scribblers, and of course pens into remote villages. In 1961 they replaced the stainless-steel ball with a carbon tungsten ball. The ball looked the same, so they designed a new logo: the Bic schoolboy, with short pants, tie, sweater, and a head that was a ball—an instant hit, and with broader appeal than Reynolds's saucy pen girl.

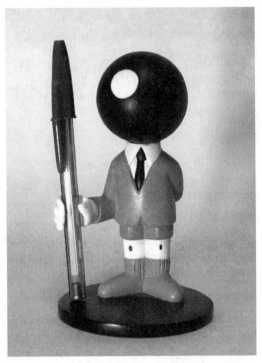

The Bic schoolboy, created to celebrate the new tungsten ball, became the face of the company worldwide. Photo by Jeff Papineau, courtesy of the author.

Yet Bich had still not cracked the American market. So with a massive publicity campaign he brought the Cristal to America— and lost a million dollars. (The first ads showed a Bic drilling through a board and still writing, but this made Bich think of a dentist's drill, so they switched to a pretty girl in a short skirt gliding on skates.) He dropped the price of the Cristal from 29¢ to

19¢ and lost another million dollars. From 1958 to 1963 he lost a million dollars a year. Bich said, "If I'd had partners they'd probably have locked me up after I lost three or four million dollars. I could never have risked their money like I risked mine." In the end he captured the market.

Unlike the genial Bíró or the perpetually glad-handing Reynolds, Bich had less social ease than his pens did. His first wife, Mimi, died in 1950, the year Bich launched the Cristal. Bich seems to have been as determined a lover as he was a businessman—his new fiancée, Laurence, had the impression of being "carried off by an ape." Domestic life had already been set: Marcel came home at lunch, never discussed business, went to bed every night at ten even if they had guests. The old servants, who didn't like Laurence, made the same soup every day, which, she says, she came to love. Bich looked like the Pope and pronounced like one, with such maxims as "The rule of nature is Competition Above All: it is the condition of survival." Actual conversation seems not to have been his forte. Laurence says he read philosophy; appreciated Cicero, Shakespeare, Racine, and Stendhal along with Joseph Kessel, Panait Istrati, and Nikos Kazantzakis; and loved *Tintin* and *Spirou*. His children dubbed him "Captain Haddock" after the irascible captain in the *Tintin* books, but they were also *"les martyrs de papa."* Despite the reading list, we get a portrait of a mind constantly churning—solving the next problem, planning the next campaign, occupied with both tiny details and the grand sweep of his business—but little given to self-reflection. Her husband was, Laurence says, attracted to lost causes, impossible challenges, like a quest for a modern Grail.

At home he faced another battle, this one with the French school system. Though his logo was a schoolboy, the French government did not want schoolchildren using his pens. Finally, in 1965, the national minister of education declared that new times required new instruments, and on September 3 issued bill no. 65-338. French intellectuals squared off in the press. Roland Barthes supported the old wooden handles with their inserted nibs (what Bich had been making up in Clichy at the end of the 1940s). Barthes scorned plastic as a "disgraced material, lost between the effusiveness of rubber and the flat hardness of metal ... flocculent ... opaque, creamy, and curdled"; plastic sounded "hollow and flat," and came in colours "only the most chemical-looking."

But the battle went beyond aesthetics; it was a war over the very character of the French citizen. Handwriting formed moral fibre. Ballpoints would undermine important practical and mental habits, the detractors claimed: the attention to minute detail, the sharp lightness with a personal trace, the taste for work that was as well presented as possible. Pundits wrangled over up and down strokes. Bich scoffed. Under the downward pressure of the hand the nib split open slightly and the stroke thickened. In the upward movements the pressure of the hand lightened so that the pen didn't snag on the paper. Thus, Bich said, "the up and downstrokes could be seen for what they were: an exalted servitude to a line of virtue. The real beauty of writing resided in the clarity and harmonious construction of the characters that the ball, more manageable, made better than the quill."

A spokesman for Waterman Pen Company in the early days had said, "Look at the example of the American Constitution. How can you obtain such a variety of beautiful signatures with a ball?" Cursive writing is at once a mode of individuality and a method of enforcing social conformity. But maybe virtue *was* an issue. One early adopter of the ballpoint was Jean Paulhan, publisher of the *Nouvelle Revue Française* and admirer of the Marquis de Sade.

Paulhan was also the author of "Happiness in Slavery," the preface to the 1954 pornographic novel *The Story of O*. He would stroll out ostentatiously with a Cristal in his pocket, to the indignation of his more conservative associates. (It would later emerge that Anne Desclos, who wrote *The Story of O* under the pseudonym Pauline Réage, conceived of it as a love letter to her lover—who was this same Jean Paulhan. Which just goes to show.)

In any case, resistance was futile. The disposable Bic came to define the ballpoint, and Italian intellectual Umberto Eco declared that, like the throwaway lighter, the Bic Cristal was an example of socialism realized. Born intentionally ugly and become beautiful because practical, economical, indestructible, organic, it annuls the notion of property and all social distinctions. Everyone, anyone has a Bic, and it's true, you don't actually think of owning them—you lose yours to someone else, you pick up another; you buy them in clumps knowing they will move hand to hand.

Bich's pen was revolutionary in a way that Bíró and Reynolds had never imagined: it was disposable. "Its low price made it an object you could throw away," said Bich, even more emphatic than Bíró about the class implications: "It distanced itself voluntarily from the de luxe finery destined to last an eternity and to be treated with respect." This was a pen you did not respect; its uniformity bordered on promiscuity: "The Cristal you threw away, abandoned, or forgot, but it multiplied to eternity and remained faithful, always within reach of the hand." Bíró's invention enabled you to have *your* pen; Bich enabled you to have *a* pen. Customers had places to go, things to do, deals to make. Leave one in the car, one on your desk, one on the kitchen table, lose one, grab another. Did anyone ever use up all the ink in their Bics? In 1950 no one had eco-guilt.

III

The subtitle of Laurence Bich's biography of her husband, *Homme de pointe*, can be translated as "point man," a man on the cutting edge. The ballpoint pen was all about being on the move and up to date, wheeling and dealing, speed and efficiency—like the smart-phone today—and it was naturally viewed with suspicion by those more conservative, and those wanting to keep them from school-children. There was the question of civic virtue, but even of personal subjectivity: the mass-produced pen produced mass-produced signa-tures. There was the whiff of sex about them—Would you want your daughter to go out with a guy who swaggered around with a Reynolds Rocket in his pocket?—and they were a mini emblem of military power, from bomber captains to captains of industry.

The "habits of contemporary life" that Laurence spoke of have changed. Perhaps we're returning to the medieval practice of signing our names in ink only once or twice in a lifetime. Now every hotel gives away pens; they're really advertising vehicles, not writing instruments, never meant to last for Reynolds's mythical fifteen years, from grade school through university. The advent of the ballpoint, an invention that must have seemed as eternal as the printing press, in fact marked the beginning of a fifty-year window of mass penmanship that is fast closing.

In 1998, four years after Marcel Bich's death, the restored Château d'Ussel was inaugurated with an exposition, "The Marvellous Adventure of Baron Bich." The creation of the pen *was* a marvel. We take the Bic ballpoint completely for granted, toss it around uncapped, use it to cut through strapping tape on packages, but in fact it's a precision instrument with fine mechanical toler-ances and a complex chemistry that took years to develop. Further, although the Cristal's supreme utility as an object obscured its beauty as a thing, it has since earned a place in the permanent collection of the Museum of Modern Art.

BEFORE LEAVING PARIS we visited the new Shakespeare & Co. opposite Notre Dame, where Hsing found Orhan Pamuk's memoir, *Istanbul*, for me. Then we caught the fast train to Geneva. It weaved through the hills, leaning like a motorcycle on banked rails, riding on magnets (Bíró's principle, imported back from Japan? I wondered) through farms with limestone knobs in them, a broad valley with tree-covered escarpments on either side, a big turquoise river beside us, white swans in placid waters. *"Bellegarde, mesdames et messieurs, bellegarde."* We had come effortlessly and quietly to Switzerland.

The next day we were in my brother Lloyd's noisy little Toyota, hurtling through tunnels at 160 kmh toward Switzerland's other edge to visit the Château d'Ussel, the crumbling fourteenth-century castle that Marcel Bich bought after he'd spent millions, and the decade of the 1970s, trying to win the America's Cup yacht race. The castle itself, a cuboid inspired by Syrian castles at the time of the Crusades, commands the valley. Appropriate, I thought, to its commanding last private owner, yet there was nothing here about Bich, only an exhibition of photographs of villages in the region. The custodian merely shrugged when I mentioned the 1998 Bich exhibition.

This was where he came from, the Val d'Aoste, a craggy zone between France and Italy known for its craggy, independent inhabitants. The old hillside houses have heavy slate tiles on their roofs; a metre square and pointed, they'd crush you if they dropped flat on you, split you in half if they came point first. This is the pass that Hannibal had lumbered through with his elephants. In 1994 the city made Marcel Bich a citizen of honour, and named him the best-known Valdôtan after Saint Anselme, who, at the abbey of Bec in Normandy, crafted his proofs for the existence of God. Bich died that year. He had come a long way from the dead-end alley in Paris.

Back in Geneva, Lloyd opened a special Bordeaux from his cellar, toasting (in my mind) our survival of the manic drive to Val d'Aoste and back. There seemed little else to celebrate. I had seen Bich's castle but I would never see Bich's factory: they had ignored my requests to visit, and when I badgered them they said no and sent me a promotional DVD. What I wanted was to see the pen in production.

"But does it have to be Bic?" said Nicole. "No?" My Swiss sister-in-law is frighteningly efficient and seems to know everyone in Geneva. "I will phone Isabelle," she said. "She knows the chief of Caran d'Ache."

The next day I found myself walking down Chemin du Foron, past the Euromaster tire service and car wash, looking for a big red pencil. A door opened on a factory that combined state-of-the-art technology, hand finishing, and machines that were almost as old as the ballpoint itself, a place that still valued craft over mass production and was embedded in the social life of the city.

The big red pencil marks the site of Caran d'Ache,
part of every Swiss schoolchild's life.
Photo courtesy of the author.

A Taste of Caran d'Ache

Every child I know has used Caran d'Ache pencil crayons. What I didn't know was that Caran d'Ache makes everything from inexpensive pencils to luxury fountain pens, markets their own line of inks, and that in Switzerland they're more a religion than a brand. The Caran d'Ache factory is a low building, just two storeys, set back from the road on grounds with big trees and flower-strewn lawns, more manor house than industrial park. Security buzzed me in and Carole Hubscher, head of promotion, came to meet me. Tall and slender, she had on elegant glasses and a business suit with a brown scarf at the neck, an ensemble that was stylish without being sexy (which of course made it so). I knew she was fluently trilingual, a graduate of Harvard, and head of her own company, Brandstorm, that promoted Calvin Klein and handled marketing for Swatch, but she was not at all daunting and had a warm laugh.

She introduced Eric Vitus, head of "R&D Couleur." He had close-cropped grey hair and wore a grey tweed sports jacket and a grey checked shirt—a surprising choice for a man in love with pigment. He smiled and said, "They call me the 'chef' because I make the recipes for the colours."

I told him of my interest in ink and mentioned that I'd learned to write with a straight pen. This was like a secret handshake: "I am from France and I too wrote with a *porte-plume!*" he said.

In the laboratory he showed me the new colours. "It is the chance I have in this company. I can use all the pigments I want! My boss and my president, they would never tell me to reduce the quantities of pigments I use." He spoke with impish enthusiasm. Then, slightly mournful, "We have developed about forty here but we will only keep a few of them." I asked if he and his colleagues ever fought about what colours they would keep.

"No, not really," he said.

"All the time," said Carole, and they both laughed.

Eric said, "Well, colour is, what do you say? Very subjective. I could have the greatest artist in the world here and that does not mean his sense of colour is better than mine." He said it lightly but his belief was fierce. "When you look at our biggest box it is one hundred twenty colours, and to develop these we used more than forty pigments. As a chemist of course I could do that with eight pigments ... but the quality—the light fastness, the purity, the brightness—this is what makes the difference." You could sense the colours teeming in his head.

I told them that I'd passed the Caran d'Ache window in the train station and a Swiss friend had told me, "My bedtime story from my grandfather used to be him telling me what was in the window."

Eric said, "Yes, it is a never-ending story; we always hear that their parents, their grandparents had Caran d'Ache pencils."

Carole said, "It is very emotional. At trade shows the people passing by, they see the name and they stop, they sit down, and they start telling us the whole story of their childhood, the memories they have, opening up the books, the smells, the colours …"

"At one point the canton of Geneva challenged Caran d'Ache and said, 'Well, maybe this year we will take another brand because you are too expensive.' You should have seen! The people almost went down into the streets; it was a real revolution. They were writing to the papers, saying, 'Look, you cannot take another brand than Caran d'Ache. We need Caran d'Ache!'"

They led me down to the factory floor, to a scene out of a Wallace and Gromit cartoon:

Pencils carried on conveyor belts up and down—*chk, chk, chk*.

Pencils pushed through a tub of coloured varnish—*shlp, shlp, shlp*.

Pencils held in a rack of one hundred and dipped once in a tray, *tch!*, to get the single dot of white paint that seals the top.

We passed a machine that looked like a giant Mixmaster. Eric said, "We glue together powders and binders, in this case it is with hot wax."

"It looks like a big pudding," I said.

"Yes, yes! The machine is from a bakery. It was used to make bread." He showed me the leads being worked—a long black string was cut into lengths, floppy still, like black liquorice, then these were put into what looked like coffee cans with holes in the sides. Finally, batches of cans went into a big dryer that turned them slowly so they would dry evenly.

"At this point there is no colour," said Eric. "If you write with it, it is too hard. So now we immerse them in a mixture of pigment and hot wax, for several hours. After that they will write."

We walked by a pile of little slats that looked like cedar shingles. One machine cut ten grooves in the slat, then another machine laid in ten leads. The next machine put glue on empty-grooved slats of cedar and then flipped them onto the leaded slats going by on a conveyer belt.

Snk, snk, snk. Pencil sandwiches.

Somewhere in the maze of old green machines the slats were split into individual pencils, ready now to be coloured. "We use seven coats of varnish," said Eric. "We are using water-based varnishes now, trying to reduce our use of solvent in everything. Also, we never use natural pigments because of their toxicity. I try to explain to people that many pigments used in the Middle Ages were toxic. Now the products are synthetic and they are not toxic."

We watched a batch of lime-green Prismalos, the water-soluble pencils, being stamped with their labels along with a number to identify the date and the batch. "If you bring me any of our pencils, even from forty years ago," said Eric, "I can tell you the date and the batch it was made from." The object itself is the record; I had not yet seen a computer.

THE ACTION of these charming machines was hypnotic, but what if you had to work with them, or become like them, forced to do the same thing all day? I watched a young woman placing instruction sheets in a tin box of pencil crayons, closing it up, putting it to the side, and then pressing on a lever to bring another rank of fifteen coloured pencils into the tin in front of her.

"We have so many different sizes that it is not efficient to do it by machine," Eric explained. Yet some of the workers were quite old, perhaps in their early seventies. I asked Carole if people tended to work here a long time. "Oh yes," she said. "We have some who have stayed over forty years."

Later I would talk to Manu Büchting, a young German editor who'd worked for two summers assembling pens in the Schneider factory in Wernigerode: "I was sitting for eight hours doing this. It was very clean and smelled of plastic. I got containers filled with the different parts of a pen, five hundred of each, and I would sit there screwing the parts together, meditating about how much money I would earn or thinking about where to travel for my work....

"Sometimes I listened to audio books so as not to become stupid, but we were actually not allowed to do that. After one container was finished, which I was looking forward to, I would get a new kind of pen to screw together. Those were the moments I thought of Sisyphus rolling the stone up the hill and then starting from the beginning again." Yet she went back for a second summer, and, "I still prefer writing with pens and remember getting blue fingers from writing a lot in high school. I liked that very much."

I was impressed, for summer jobs can engender lifetime loathings. After two months as a grill cook at the KFC in Banff, for the next five years I couldn't eat even my mother's fried chicken; the smell of those "secret spices" still makes me gag. The next year I worked on the assembly line in a window factory (with no windows so that the workers wouldn't be distracted), putting on the aluminum channels that surrounded the two panes and then, with evil-smelling solvent, cleaning off the glue that had oozed up onto the glass. The cleaning product is probably illegal now. I spent the first two weeks happily stoned until I got used to it.

As we moved into the next room we passed a woman hosing out plastic containers the size of big garbage cans. "She is cleaning," said Carole. "This takes up almost half our time, the cleaning. It is very expensive." Admirably ecological, and part of their devotion to green production, but I wondered what it was like for the woman who cleaned. It's so easy to forget the human labour, the cost to the body, behind these things.

WE MOVED FROM pencil crayons to mechanical pencils ("We invented the clutch pencil in 1929," said Carole, handing me one so that I could see the three little prongs that did indeed clutch the lead), then we reached the ballpoint pen section. In a side room a machine that looked like a lie detector—or Bíró's modified record player—was making rows of circles, elegant Slinkies to infinity. I told Eric they should frame it as art and charge money for it. Kidding, he handed me a chunk from the garbage pail, but when I said I meant to keep it he said, "No, no!," tore me off a fresh piece, and folded it carefully for me.

I told him about the problems Bíró and Bich both had in finding the right balls. He smiled. "We make our own points right here in the factory," he said, handing me a little cylinder of metal and then the ballpoint nib that it would become. "Everything is made here and everything is controlled many times [*contrôler* = to check]."

Kids would always want coloured pencils, but what about pens, were they going to disappear? Carole said, "Of course there is this phenomenon of the computer, but there is a comeback to real values, to using the hand.... When you use a pen you're making your sentence in your head first, thinking how you are going to build up your text." (The Caran d'Ache website quotes *Don Quixote*: "The pen is the interpreter of the soul: what one thinks, the other expresses.")

Also, the consumer had changed. "Before it was just getting things very quickly and throwing it away. Now it is not the same." The era of the throwaway had passed; these pens were un-Bics, designed to last many years, through many refills. Eric gave me a handful of parts for a single pen, all metal.

But what about ink? I told Eric of Bíró's difficulties, how every time the humidity changed he had problems. "We know! We have the same problems!" said Eric, and laughed. "And the technology is important … the ball … the diameter … you have very thin channels. But it is very rare now, because we control the viscosity of the ink."

We looked into a room with two young women working at microscopes. Eric said, "Here she is controlling the ballpoint ink. You see she puts it on a slide under the microscope to look at the crystals. If the crystals are too big this can cause problems. Usually we have none, but we want to be sure."

WE PASSED THROUGH another door, and suddenly everything was quiet. Carole said, "This is the luxury pen division." I knew from their catalogue that the barrels they were working on were plated with silver, and gold, and rhodium.

"The ballpoint is complicated, yes," said Eric, "but it is also very difficult to make a good nib for the *plume* [fountain pen]. We make many controls." The workers looked up—clearly we were intruding; much of the work is done by hand. Through a window in another door I could see a man in a white labcoat with black gloves on. "We are one of the few who use Chinese lacquer," said Eric. "This man has been working here maybe thirty, thirty-five years. And it is very specific, the Chinese lacquer. We buy the raw lake [resin] but we prepare it for many, many weeks."

"Maybe I should buy a lacquer pen," I said. "Are they very expensive?" Eric and Carole laughed. "Oh yes," they said.

"So," I pressed, "how much?"

"Well they are not expensive for the work that goes in," said Carole. "But …" It was clear that it was not for me. They didn't want to embarrass me by naming a figure.

"More than a thousand francs?" I said, as if that were not too much (around one thousand Canadian dollars, which was way too much).

"Oh yes, much more," said Carole, and we moved on.

I remarked that they must have a whole division to come up with the variations on the colour names for their 120 pencil crayons (Opaline Green, Veronese Green, Spruce Green), and they laughed. More seriously, I asked why they had only nine colours of ink—Blue Night (dark blue), Blue Sky (medium blue), Storm (purple blue), Caribbean Sea (turquoise), Carbon (black), Amazon (green), Saffron (yellow), Grand Canyon (brown), Sunset (red). Carole said, "We have tried to match the colour of the ink with the colour of the pen, but you are selling blue and black, basically. Also, with pencil leads you can keep them ten years, twenty years … The problem with liquid products is that we need to be sure they will leave the company before one year."

When I asked if they made their own inks, Eric looked uncomfortable. "We make our own recipes." I didn't press him. It's very important to them that their products are Swiss-made. On their sign out front, beside the big pencil, they proclaim (in the slightly Anglicized French familiar to Canadians):

PARTOUT DANS LE MONDE
CARAN D'ACHE PORTE LES VALEURS
DU SWISS MADE

And again, under their logo:

SWISS MADE PAR NATURE

Carole had said, "Swiss-made still has a value abroad, I think. A Swiss-made watch is something different I think, and they are ready to pay the premium of a Swiss-made brand. Maybe not in North America where price is the first issue, but for us. The quality is there. With a Caran d'Ache pencil you can drop it as many times as you like and it won't break." Had my question about the ink source opened up one of those fissures that lead from the innocuous surface to something more fraught underneath? The friend who'd been charmed by her grandfather's stories of the Caran d'Ache window had said, "Yes, I heard that everything is made in Switzerland—and by Swiss people too." An important distinction for her.

In the Valais, a conservative canton, I'd seen a poster showing a group of white sheep kicking a black sheep over a fence—the sheep were cute, but it was about kicking foreign workers out. Was "Swiss-made" code for "No immigrant workers have touched your pen"? Yet the company website celebrates the complex lineage of its name: it was inspired by the French cartoonist Emmanuel Poiré, who took the pseudonym Caran d'Ache, which comes from the Turkish *karan tash*, literally "black stone," the word for graphite; and in Russian, *karandache* means pencil. As with Bíró's Jewish heritage in Hungary, I felt out of my depth, that there were shifting currents underneath. Was I an innocent abroad or paranoid policer of political correctness? Freud said that some-times a cigar is just a cigar, but I was coming to feel that a pen was never just a pen, that it was always a signifier enmeshed in a web of social implications.

"No," said my brother Lloyd, dismissing my attempt to elevate blatant consumerism. He's thrifty, and I'd come home with a $500 fountain pen. They'd given me the staff discount. It was the Leman series (for Lake Geneva: Lac Leman): it had a beautiful fluted silver cap with *Caran d'Ache* etched into the silver around the base, and a nib with looped lines that reminded me of a Lawren Harris mountain. The top was heavy and the blue of the barrel as deep as the lake for which it was named. Yes, it was over five hundred francs, but, I reminded myself, so much cheaper than the Chinese lacquer pen. It was a cultural artifact; it would be like joining a family. I'd also bought an orange one, because how to justify a five-hundred-franc pen to your wife except by bringing her one, too? I'd clutched my pens and walked to the tram in a daze, shocked that I'd spent so much, delighted that I had.

Back in my brother's kitchen, I spread my prizes on the table: little packets of pencil crayons for the children, and then the dark-blue boxes that opened up to reveal (like silk-lined coffins—I've always found pen boxes a bit creepy) the fountain pens lying in state within.

Nicole, my thrifty Swiss sister-in-law, said, *"Combien?"*

I told her. She pursed her lips.

"Hein."

I told Nicole I had a new theory: "I have figured out what unites Switzerland." (Lloyd claims there's no such thing as Switzerland—there's the French section, the Italian, the German. Then you have the divisions within cantons, even within valleys. In the Valais, where Nicole's family has a chalet, there are the *haute* and *bas* Valais. The hill dwellers above sneer at the soft city dwellers in Martigny below, who return the favour by scorning the sheep-herding peasants above.) "It's not chocolate, it's not Swiss watches—it's Caran d'Ache. You grow up with their pencil crayons, you use their pens as an adult, you tell your children stories about their advertising displays …"

"Oh no," she said, "you are wrong. Because Suchard choc-olates—you know the one with the cow?—they are known to everyone. And Lindt chocolates from the German part. And Toblerone. And Nestlé, well it is a big company now, but we remember it when it was small. And the watches—each one is made by a family. Caran d'Ache is not the name of a family!"

My point exactly. There are many watchmakers and every

chocolate devotee fiercely champions her own brand, splintering the country again. I didn't say that, for I also had a theory about Nicole: that if you present an idea she wishes *she* had thought of first she'll argue you into the ground to prove it wrong. So I took her opposition as validation and compliment. And I did not say that for Christmas last year she had given me a Caran d'Ache ballpoint. What I did say was "It's a perfect size. I think you cannot find such a pen in Canada." I reminded her of how the night before she'd shown me the Prismalo coloured pencils that you can dip in water to get a kind of wash. She'd said, "You are not supposed to, but of course we used to put them on our tongue. I remember very well the taste of Caran d'Ache." Then she laughed, and licked the pencil. "It is like Proust." It transports you to your youth.

PRINTERS' INK

The printing press bed, with type in the chase, secured by a quoin.
Photo courtesy of the author.

Grinding the Bones

I

Every schoolchild knows that Johannes Gutenberg revolutionized the Western world by inventing the printing press. But every schoolchild is wrong—printing presses had been in use long before, not only for block printing but with movable metal type. The Koreans had been doing it for almost a century, possibly two, depending on whose account you believe. The real breakthrough was an ink to work with that type. There are two basic forms of ink—writing ink and printing ink—and they really should have different names because they're as different as syrup is from water.

The Chinese ink used for woodblock printing was virtually identical to the ink used for writing—thin, water-based fluid that, as ink historian C.H. Bloy says, "did not spread evenly over the letters, but stayed in globules." Gutenberg's ink, on the other hand, was oil based. He probably got the idea from painters, who were working with a new medium: oil paints, perfected by Flemish painter Jan van Eyck in the years Gutenberg was working on his press. In the late Middle Ages, printers and painters were members of the same guild; painters were involved in manuscript illumination and worked closely with engravers. Gutenberg, a goldsmith by trade, would have been aware of developments in the field and known about the drying oils in varnishes, so the materials for his ink were at hand when the need arose. Just as master painters made their own colours, Gutenberg was his own ink maker. I knew that the ballpoint pen used a variation of printers' ink. Now I wanted to go back to Gutenberg, to the ink that started it all. I expected the quest would take me to the birthplace of the press in Mainz, Germany, but instead the ink trail led me to Texas and then to a ranch outside Cove Fort in the Utah desert.

It is a truism that the arts do not follow the linear advancement of the sciences: Einstein trumps Newton but James Joyce does not supplant Shakespeare. Contemporary printing presses are infinitely more sophisticated than Gutenberg's converted wine press, yet the printing itself hasn't surpassed the work done on Gutenberg's Bible. At the Gutenberg museum in Mainz I'd looked at the gorgeous Bibles in their glass cases, but I had nothing to compare them with, and though I tried hard to feel thrilled and reverent, I kept thinking about the vaults we were standing in and whether a power failure would trigger an automatic closure and we'd all be trapped. At the Harry Ransom Center in Austin I'd looked at two editions of the First Folio of Shakespeare's plays: their ink was pale, uneven, and grey, and they reminded me of my old Penguin paperbacks. Then I went downstairs to look at the Gutenberg Bible: a century and a half older than Shakespeare and the ink was rich, deep, and glossy. It looked as if it had just come off the press.

As C.H. Bloy says, "In the works of Gutenberg, Fust and Schoeffer, and, indeed, in nearly all incunables (works printed in the first fifty years after Gutenberg, Latin for the "cradle"), the eye is immediately struck by the blackness of the ink, without the unpleasant haloes round each letter that one associates with printing of the seventeenth and eighteenth centuries." I finally understood why people made such a fuss about Gutenberg: his work was better than anything else that had been produced for decades afterward. In some ways it had never been equalled.

"INK?" SAID MICHAEL WINSHIP, when I ran into him and his book-history students coming out of the Ransom Center in Austin. "Then you have to talk to Steven Smith. He teaches a book arts program up in College Station. He's always building things, setting fire to things. Give him a call."

I did, discovered they'd be making printers' ink that Thursday, and accepted Steven's invitation to come up for some ink and some barbecue. I found the Cushing Library on the nearly deserted Texas A&M University campus, and Steven Smith strode out to meet me. He looked younger than his photographs, in his mid-forties I guessed, though it was difficult to tell because he fell into the enthusiastic librarian category (perennial kids, who retain their excitement about books, as opposed to the snarling librarians who hate people who disturb the perfection of the tidy shelf). We sat in a big boardroom for a few minutes and I talked about my project, how I didn't really know anything about printers' ink, but had read Bloy. The mention of Bloy—"I was going to recommend that," Steven said—seemed to consolidate my bona fides.

In London in 1967, as the Beatles' "Sgt. Pepper's Lonely Hearts Club Band" climbed the charts, U.S. bombers attacked Hanoi, and Expo 67 opened in Montreal, C. (Colin) H. Bloy published a little book called *A History of Printing Ink: Balls and Rollers 1440–1850*. A bare hundred pages of text with an appendix of recipes, the book comes from a different universe from Abbey Road Studios. My former self, who wanted a Gibson Les Paul guitar and thousands of adoring nubile fans, would have scorned it. Now the book fascinated me. I worried about becoming a complete ink nerd, but felt proud when Adam Gaumont, the editor of the campus newspaper, came to my Book History class early one day and said, "Have you read Bloy? He's terrific!"

Upstairs, groups of students were huddled around hand presses, one inking with the brayer (a thing like a thick rubber spatula), one putting in the paper, one pulling down on the handle, one taking out the completed sheets. Steven gave me a labcoat and took me over to a small table where he'd placed a mortar and pestle with black powder. Behind the table on the shelves he showed me the wooden things they use for the ink balls. "These are ball stocks. Moxon says they're usually made out of alder or maple." They looked like goblets without a base, the cup about six or seven inches in diameter. Steven pointed out the wool that goes in them, the combs used to tease the wool, and the leather that covers the wool. Ready, they looked like boxing gloves with handles. "Printers always soaked the leather overnight. It's sheepskin, and you use a sheep's foot as a hammer to tap in the tacks that hold the leather on the ball." They cheated and used a real hammer.

He showed me jars of varnish. "This one's from a couple of years ago when Steve Pratt and I made it on the loading dock at the university. It's supposed to spontaneously combust—you need that to get the varnish the right thickness—but we turned away for a moment and suddenly there were twelve-foot flames going up the side of the building. Instead of going for the fire marshal Steve was running around trying to find a camera. So we don't do that here anymore. After lunch I'll be showing a video of ink making with Steve in my backyard." He took down an oblong plastic container, about the size of a loaf of bread. "Here's our pigment—carbonized bone," he said, showing me the broken chunks that looked like dark spare ribs.

"So, what kind of bones are they?" I said.

"Sheep or goat. Pratt's got a farm up in Utah so he's always got something dying on him," Steven said. "He chars them on the barbecue. If you're doing it you have to remember to close the vents to cut off the oxygen—that's why they don't just burn up." I didn't think I'd be doing that, but it was good to know. Bloy tells of Hoyau, an ink maker in Paris who ran afoul of the law in 1732 for burning sheep bones with lees of wine. The "sentence de police" states that this "causes such a great infection in the area that those people who are healthy, have difficulty in putting up with this odour, which is much worse for sick people." Further, "In order to make this ink he is obliged to have a large fire, and the smoke mixing with the smell enters the apartments of the said citizens, in spite of the fact that they close all possible entry points." The neighbours hated the smell, but more than that they feared Hoyau would burn down the neighbourhood. The police sided with the neighbours, and "the said Hoyau" was "required to withdraw to a distant place to there burn the said sheeps' bones and lees"—or face a fine of a hundred livres, the equivalent of a pound of silver. I wondered if the said Steve Pratt received any grief for his bone burning. Would his wife object to the family barbecue being used for projects?

Smith had turned back to the table and the mortar and pestle. "If you want you can grind that some more." I stared at the pile. These were real bones from real sheep. I love lamb chops, but there was something about grinding the bone of a living thing that seemed a desecration. I'd talked with Jim Stroud, the ink authority at the Ransom Center: a tall, chain-smoking Texan who at hourly intervals would go out and sit under the live oaks in front of the

Center with his Camels and think. He told me about grinding ink at his ink camp in south Texas: "You can imagine trying to grind up bone. It's like grinding up a cremated body. It's *hard*. And then if you're using a brush for calligraphy, one little piece of grit would ruin it. The particles are so much finer in lampblack. That was probably the attraction of lampblack inks: no grinding issues." I knew about lampblack from Cennino Cennini, a Renaissance painter, who says it's simple to make: just burn linseed oil in a lamp, place a "good clean baking dish" over the flame, and then carefully brush off the soot.

I LOVE OLD TRADE MANUALS—there's something in the prose that takes you close to the thing itself, as when Seamus Heaney describes a spade and you feel it in your hand. The author had to write in a way that would bring an object before you, enable you to assemble it, fix it, distinguish it from something else almost the same. There were no instructional videos; the writing was itself a craft.

Jim Stroud said he sometimes goes back to the old handbooks over more modern works. He's a big fan of the learned and literary C. Ainsworth Mitchell (he speaks of gallnuts as "curious vegetable excrescences"), whose *Inks: Their Composition and Manufacture*, first published in 1916, is still cited today. We agreed that nobody is going to read one of today's computer instruction manuals in a hundred years. Or five hundred years. Cennini's *Il Libro dell'Arte* (*The Craftsman's Handbook*) is still in print in an inexpensive Dover edition, not bad for a work written around 1400, fifty years before print existed in Europe.

Part of the magic of early technical writing is that it artlessly reveals the life around it. As you read one of Cennini's recipes, a Breughel painting takes shape in your mind: someone pulverizing red lead, mixing it up with the resin, frankincense, and the rest while the oil is being brought to a boil on the fire. Then stirring for two hours, testing the cloth, finding—damn!—that the black isn't black enough so he has to burn some vine branches in a pot, grind and dry them, and then add them in to the mix. I could see why people love to read old cookbooks. Cennini's instructions are embedded in the natural cycle. He tells you to make goat glue "in March or January, during those strong fronts or winds"; to do your gilding in "mild, damp weather"; to apply the final varnish on windless days after you have "warmed your panel in the sun." There were no humidified and temperature-controlled labs to work in; you produced your materials at the appropriate time of the year; each stage in the process had its own season. (Even Mitchell, writing at the beginning of the twentieth century, notes that the best season for the manufacture of lampblack is at the end of autumn or beginning of winter.)

Cennini's world was also deeply social. He talks about getting recipes for red ink from the friars, glue and pine rosin from the druggist, charcoal from the bakers, cuttle to rub down the panel from the goldsmiths, miniver tails from the furriers. He goes out into the countryside with his father, digging out seams of ochre: "These colours showed up in this earth just the way a wrinkle shows in the face of a man or woman." Women are much on his mind. "There is another cause which, if you indulge it, can make your hand so unsteady that it will waver more, and flutter far more, than leaves do in the wind, and this is indulging too much in the company of woman." Nonetheless, working the lapis lazuli (for the "most perfect" blue used for the Virgin Mary) "is an occupation for pretty girls rather than for men; for they are always at home, and reliable, and they have more dainty hands. Just beware of old women." He never does specify, but you sense household tensions.

My small pile was already finer than sugar, but it had some slightly larger particles, and I had to use all my force against the side of the mortar to shatter these fragments into dust. It was hard work, much harder than grinding gallnuts or gum arabic, and I didn't like the brittle crackle that reminded me this wasn't stone but bone.

"You know, I think that's good enough," said Steven beside me. "Let's go ahead and mix it up with the varnish." He poured the powder out of the mortar into a neat pile, about three inches across and an inch high, on the polished stone slab. Then he took the lid off a mason jar of varnish and upended it over the pile. Nothing happened. Then slowly, slowly, like thick maple syrup that you've left in the fridge and forgotten to warm, the dark varnish started to ooze out over the lip of one side of the jar. It hung there, gradually getting bigger, like a possum hanging upside down, until finally its weight was too much to bear and it stretched out in a line, dropping down to the pile of black bone. Dark amber in its pour, the varnish became a glossy black on the powdered bone. "Beautiful," I said.

"Isn't it?" said Steven. "Now we'll mix it up." He took a putty knife and spread the varnish around, trying not to scatter the dry powder as he worked the varnish, turning it over until he had it all wetted. "Steve Pratt says the best proportions are 50/50, but I think that's too much pigment. I think 60/40 or even 75/25 is better. Look how thick this is." He upended the mason jar and the varnish came a little more quickly this time, drizzling down onto the top of the pile. Steven worked the thick liquid in, churning hard with his wrist. "That's better. Now if you can take this—" He pointed to a heavy iron cylinder about eight inches high. "That's our muller. Steve made it out of some old vehicle part. Maybe a drive shaft from a tractor. You have to have something heavy, it doesn't matter what it is. So take that and just start working the ink, mixing the pigment in."

Now that the powder was in the varnish I didn't think of it as bones anymore; it was the pigment, and what I had, Steven told me, was a colloid. Unlike solutions where one substance is dissolved in another, colloids are mixtures where one substance is dispersed evenly throughout another. That was my job—to disperse evenly. In a colloid (the word means "glue-like" in Greek) the particles stay particles, and have a diameter of between about five and two hundred nanometres—one billionth of a metre. I didn't think we were going to get down to two hundred nanometres today.

THE FIRST RECORD of an independent ink maker is Guillaume de Luanay in Paris in 1522, only seventy-five years after Gutenberg. But the ink industry didn't develop until the 1700s, and even then the major issue was the amount of physical effort required to grind pigments: the inks were very coarse. Cennini had told his apprentices that you can never grind too much: "Take some clear river or fountain or well water, and grind this black for the space of half an hour, or an hour, or as long as you like; but know that if you were to work it up for a year it would be so much the blacker and better a colour." You could get your "Printers black" from the salters—dealers in salt—but it was still ground by hand. It wasn't until the 1820s, three and a half centuries after Gutenberg, that Frederick Koenig, the inventor of high-speed printing presses, developed a power grinder. Ink historians claim that the real breakthrough in commercial ink came not through any chemical formula but with the invention of the industrial grinder. Historians of anything are always saying "The *real* turning point in human history came with the invention of the _____," but after grinding away at the bones I was prepared to put the power ink grinder up there with Velcro.

The other big advantage to buying your ink, aside from not having to grind your wrists to powder, was that it saved you the risk of burning your house down. As Bloy says, "No longer did the printer have to spend time and trouble in making his ink, a tiresome and dangerous process ... as the oil was liable to catch fire and run over the brim of the kettle in a stream of liquid fire, a threat to the mainly wooden habitations of the time." Yet something was lost: precisely because of that risk, ink making had become a social activity. One Theodore Goebl, looking back near the end of the nineteenth century, recalls how twice a year the printer and his workers would travel outside the city walls to make ink, eat *abgekröschte* rolls, and drink schnapps. Goebl remembers how the city wall was blackened by the pot of flaming oil, and one can imagine the drunken apprentices lurching about, burning their fingers as they grabbed the rolls out of the oil. The bread was supposed to make for smoother boiling, or maybe it was a way of estimating the temperature of the oil, or maybe it was just how the varnish maker made lunch. No one is quite sure. Everyone *is* sure that linseed oil is good for you. It was used for treating cuts, bruises, and burns; it was supposed to help tuberculosis; the oil that ran down the screw of the press into the pan was useful for hemorrhoids. Bloy tells of an aged varnish maker who claimed his longevity derived from drinking a cup of raw linseed oil every day; apparently it can lower cholesterol, says Bloy, but physicians I have consulted advise against it.

Steven was heating oil in a slow cooker ("It never gets hot enough to catch on fire") to toast buns later, in the tradition of the *abgekröschte* rolls.

"We use flaxseed oil now. Organic flaxseed."

"How much did that bottle cost you?" said Chris, one of the assistants.

"Twenty bucks. And we used to buy a whole gallon can of linseed oil for a quarter of that, but people were worried because it says on it 'the State of California.' Yeah, see, *California*"—he said it with Texan contempt for those effete eco-obsessives on the Coast—"'has determined that it may contain elements that contribute to cancer.'" I said it was probably those organic flaxseed producers who wanted to stop people buying cheap linseed oil—though maybe the processing of linseed oil didn't have the same standards as the organic flaxseed used for food. I told him how in grad school I'd used paper towels for coffee filters until somebody told me they didn't have to meet the same standards as a food product and were full of all kinds of bleaches and things.

He agreed that was probably it, but said, "Hey, I ate the *abgekröschte* rolls for several years running and I'm still here." Linseed oil, contained in the seeds of the flax plant, was high in price and so couldn't be used in the cheap newspaper inks, but as Philip Ruxton says, it was "unquestionably the best vehicle for the better grades." The chief virtue of linseed oil is that on exposure to air "it dries rapidly to a hard surface, which adheres very firmly to the paper, and is not readily affected by further exposure to light and air."

Ruxton's book was another little gem that I'd stumbled upon after Bloy. Everything about Ruxton was succinct and precise, and I liked his phrase for printers' ink: a "glutinous adhesive mass." The syllables moved as slowly as the varnish oozing out of the jar. Produced in 1918 by the Committee on Education of the United Typothetae of America in Chicago (*typothetae* was a made-up word from Latin and Greek, created for this association in the early nineteenth century to make master printers sound even more lofty), Ruxton's book was part of a series of sixty-four pamphlets intended to educate typographic apprentices. It included a hundred study questions in the back, ranging from fundamentals (#1 "What are the two ingredients of all printing ink?") to crafts (#27 "What is bone black and how is it made?") to shop practice (#78 "How can waste of ink by skinning be prevented?"). Answers: #1 Carbon and oil; #27 Charred bones that are ground ("It can never reach the same degree of fineness as lampblack"); #78 When you open a can of ink do not dig down in the centre, but take from the surface and then cover the ink again with the paper disc. Other pamphlets in the series covered everything from Typesetting to Estimating and Selling, to Health, Sanitation, and Safety. I wondered if these pamphlets were actually successful in teaching apprentices.

STEVEN WAS SHOVELLING the mass in from the side with his putty knife. I turned the muller, leaning into it, trying to mash the particles into the varnish. The mixture flattened out matte, with little flecks in it. Where I'd passed it oozed back up to a glossy mound that looked like perfect ink, but when I squished it down again the flecks always reappeared. Todd, another assistant, came over.

"Todd's our master printer," said Steven cheerfully.

"Looks grainy," Todd said grimly, and walked away. We kept working.

Twenty minutes later I was still turning the muller. "This reminds me of those passages in Cennini where he tells the apprentices they could grind to the middle of next year and it still wouldn't be fine enough," I said.

"I think it's getting better," said Steven. "I think we're almost there." The other assistant came over. "This is Chris, he's our finest compositor," said Steven cheerfully.

"Looks grainy," Chris said grimly, and walked away. I remembered how Cennini says "Grind as much as ever you can stand grinding them," and said to Steven that if you were an apprentice in a Renaissance workshop this is what you'd be doing for the first whole year. It was hard work, and if I didn't keep telling myself that it was like being a Renaissance apprentice I'd have to admit that it was boring. Plus, I didn't seem to be making any progress at all. Goddamn sheep bones. I leaned into it, twisting the muller into the muck. One of the young students came up. "I just wanted to ask if we can … Oh, is it supposed to be all grainy like that?"

"Not really," said Steven. "I think we started out trying to make too big a batch." He answered her question and she left. "Everybody's a critic," he said.

If I ever did this again I would really pulverize the sheep bones first. Smash them with a hammer. Before combining them with the varnish I'd crush them to powder, reduce them to dust. The ink—for that's what it now was, however gritty—kept oozing up into lovely deceitful mounds that looked smooth but when flattened revealed their flecks. Bloy said that bad grinding could lead to "monks" and "friars" (black spots on the paper and white spots on the letters); he said the "undispersed lumps would choke small letters and cling to the sides"; he also said "the physical effort of grinding an ink with a muller on a stone was considerable and there was no substitute for elbow grease." Thanks, Bloy.

"You know, I think it's as good as the ink we made last year," said Steven, and reached for a little bottle on the shelf. He poured off the water on top that he'd used to keep it sealed and ladled some out for comparison. "Let's try it." He spread it on the ink slab. Cennini talks about his porphyry slab (an Egyptian rock with feldspar crystals, used for triturating—grinding—drugs), and I asked if this was the same. "I doubt it. I picked that up at Home Depot—but we have a nice piece of marble over there."

I glanced at a pile of what looked like paper baskets and noticed that a couple of the students had them on their heads. "Printers' hats," said Steven. "You'd make a fresh one each morning with one of the new sheets that had just come off the press. It was like hair nets in a kitchen—you had to keep hair out of the press. Just a single strand on the type would ruin the page. You don't have to wear one, though."

Steven led me over to the press. With Todd hovering by to make sure I didn't wreck his baby, I was going to be allowed to print a page. I raised the frisket—a metal frame with a windowed cut-out of oiled paper that covers the space between the type and the edge of the page to be printed, so that the paper stays straight and remains clean—and positioned the paper on the tympan. Then I folded the frisket back down, flipped the tympan over the type, and slid the press bed under the platen—the heavy plate that presses the paper onto the type. Todd was trying to be patient. His instructions were clear, but I was having trouble keeping the terms straight (*Okay, the frisket is frisky, the tympan is thin like the head of a drum, the platen is a plate that goes splat*). Everything was different from a roller press where, as on a typewriter, the platen is the cylinder that holds the paper. At least the block of type was the same—the *forme* (type with the wooden blocks of "furniture" to centre it), in a *chase* (the metal case), secured with a *quoin* (the expanding mechanical device, tightened with a lever and a key, to put pressure on the chase, holding everything in position). I broke into a dozen flustered movements what was for Todd one smooth continuous action, but finally I was ready to print.

I hauled back on the bar with both hands to bring the platen down, putting my whole body into it, leaning out like a wind-surfer. There was even a slanted block underneath the press to brace your foot against. (In the *Encylopédie* illustration the printer on the far left has the frisket flipped up and is arranging the paper on the tympan. His partner is applying ink with the ball stocks. The printer on the right, though he's pulling one-handed, is using the block.) I pulled once and the platen came down; I lifted off, slid the bed in again, and gave a second pull to do the bottom half

Early printers at work: on the left, one applies the ink while the other positions the paper on the tympan. Photo by Jeff Papineau from the original in Diderot's Encyclopédie, *(1769), Bruce Peel Special Collections, University of Alberta.*

of the sheet. When I let the lever go it swung back out of the way to the stopper on the other side.

"That's the best part, isn't it?" said Todd. "So satisfying, and it probably saves only two seconds." He laughed. "But for those commercial printers, each second saved made a difference. The best ones apparently could do 240 sheets an hour."

I thought for a minute. "That's one every fifteen seconds. Including lifting the frisket, positioning the sheet on the pins, closing the tympan, sliding the bed forward, hauling on the lever to press down the platen to squish the paper onto the type…."

"And don't forget the dwell," Steven said. "You're supposed to hold it for just a little bit at the end of the pull, before you release." He went on, "I'm not so sure about those numbers. I've read a thousand a day, which seems more reasonable." Still, the numbers were staggering. They worked hard, these guys. After half a dozen pulls, I thought to myself that I wouldn't have wanted to pick a

fight with a printer. I'm sure you'd get efficient at it, but I could feel
it in my biceps and my lower back. Think of doing pull-downs on
the weight machine at the gym for twelve hours a day. How many
pulls for one Gutenberg Bible? An old horseback-riding manual
rightly observes: "Riding, like every other corporeal exercise, does
not feel like what it looks like." But it also looks much different
once you've experienced it.

*Dwell and release, from a 1628 engraving of the bindery of Laurens
Koster and the press that allegedly pre-dated Gutenberg's.
Photo by Jeff Papineau of engraving by Scriverius, from a reprint of
Moxon's Mechanick Exercises on the whole Art of Printing.*

ANY QUESTION I ASKED Todd or Chris usually generated an answer that began, "Well, Moxon says ..." He sounded like a contemporary, some guy who lived up the road in Madisonville. They would often mention Moxon and Pratt in the same breath. But Steve Pratt was alive and well in Utah; Joseph Moxon was born in 1627, eleven years after Shakespeare's death.

Young Moxon began learning the printers' trade at age nine in Holland, where he worked with his father (who'd been reported for printing libels against the English government). By the time he was twenty they were back in London, producing books together. These were turbulent times—civil war between supporters of Cromwell and Charles I had been raging since Moxon was fifteen (1642). Over the next five years, while Cromwell came to power, placed the king on trial, and then beheaded him, Moxon was falling in love, marrying, and with his wife Susan expecting their first child, a daughter. In 1655, while Cromwell was dissolving Parliament and prohibiting Anglican services, Moxon was finishing his book on architecture, *Vignola, or the Compleat Architect, shewing in a plain and easie way, the Rules of the Five Orders in Architecture*, which included an Alphabetical Table of all "hard words" for the reader "disheartened" by the strange terms used in the text—Moxon was always considerate of the non-specialist. Finally Moxon achieved security: in 1660 Charles II returned from France, and within two years he appointed Moxon royal hydrographer, charged with making globes, maps, and sea charts.

The stability did not last. Susan died and the Great Fire of London destroyed his print shop. He remarried (eighteen-year-old Hannah Cooke; he was thirty-five), and ten years later we find him in arrears in his rent: "Hee was aground ... apprehended in a low condition," according to the Index of Deeds. However, Moxon had been moving in the right circles. He drank coffee with Samuel

Pepys, who through the 1660s was writing the diary that would make him famous, and with Robert Hooke, surveyor to the City of London after the Great Fire. Hooke was the curator of experiments at the Royal Society, and though famously irascible (he quarrelled with Isaac Newton over credit for work on gravity, and his biographer describes him as "melancholy, mistrustful, and jealous"), he took a liking to Moxon. Fascinated by the practical side of things—he invented the pocket-watch spring—he championed Moxon's writing on the mechanics of printing. He would have approved Moxon's caution that "the bare reading of these Exercises" wasn't enough, that "the Cunning or Sleight or Craft of the Hand cannot be taught by words, but is only gained by Practice and Exercise."

In January 1678, likely with Hooke's influence, Moxon was elected a Fellow of the Royal Society, the first tradesman to be so honoured; in that month he began to issue *Mechanick Exercises* in monthly instalments. The *Mechanick Exercises on the whole Art of Printing* was a groundbreaking work, the first printing manual in any language, but sales were disappointing. As the modern editors of the work point out, class was an issue: on the one hand there were few working men who felt they could learn their craft from a book, and on the other hand there were few "gentlemen even among the members of the Royal Society who would feel that such practical handbooks should have a place in their libraries." Moxon would have been pleased to have the head of a university rare books program, three hundred years later, speaking of his work with such respect and affection.

Most manuals I've encountered, whether on fixing your bicycle or setting up your computer, assume inside knowledge. They skip some basic, crucial step because they consider it too logical or intuitive to be mentioned. When my children were small I made muffins. Every cookbook warned that you must combine the wet and dry ingredients quickly; stir too many times and you'll flatten out all the air pockets and get leaden muffins. I made batch after batch with ever more acrobatic flailings to try to wet the ingredients swiftly, and still I was feeding my children hockey pucks with raisins. (When I was old enough they told me they threw them away at school; they didn't want to hurt my feelings.) Then I came across a cookbook that told me the obvious: use a small bowl. If you have a big one you have to move everything around too much. I wanted to kiss the editors on both cheeks. Moxon would have told you that. He omits nothing. He doesn't just tell you how to take a sheet of paper from the heap, he gives you the choreography: the pressman "nimbly twists the upper part of his Body a little backwards ... drawing the back-side of the Nail of his right Thumb on his Right Hand nimbly over almost the whole length of the Heap, the better to see he takes off but one sheet."

Moxon insists that if the ink is bad the quality of the type, the beauty of the typeface, count for nothing. Ink determines the final impression. We all know this, having written brilliant letters on elegant letterhead and seen what a fax machine reduces them to; it seems to alter the very quality of the prose. The key, Moxon says, is good linseed oil: the Dutch make their varnish "all of good old *Linseed-Oyl* alone," but to save money, the English ink makers "mingle many times *Trane-Oyl* ... which *Trane-Oyl* by its grossness, Furs and Choaks up a *Form*, and by its fatness hinders the *Inck* from drying; so that [the print is] dull, smeary and unpleasant to the Eye."

I could tell he'd like to take my inck out back and slap it upside the head.

If your wrists still ache from turning the muller, you'll find passages that you'd otherwise have glossed over now leaping out at you. Such as this one, in which Moxon is lamenting how the

English printers, unlike the Hollanders, cut corners with their ink: "They to save the *Press-man* the labour of *Rubbing* the *Blacking* into *Varnish* on the *Inck-Block*, *Boyl* the *Blacking* in the *Varnish* … which so *Burns* and *Rubifies* the *Blacking*, that it loses much of its brisk and vivid black complexion." I had *Rubbed* the *Blacking* into *Varnish* on the *Inck-Block* for hours, and if I'd known about throwing the blacking into the *Boyling-hot* ink I would have done so in a trice and the hell with the "brisk complexion." My sympathies were all with those deadbeat shortcutting Englishmen, not the fastidious Dutch. Which is why my ancestors produced printing so much inferior to their compatriots in Holland. Everybody found ink making a chore, "laborious to the body … noysom and ungrateful to the sence," so the lazy "English master-printers do generally discharge themselves of that trouble; and instead of having good inck, content themselves that they pay an inck maker for good inck," which, cautions Moxon, may "yet be better or worse according to the conscience of the inck maker." I knew from Bloy about how little conscience that was.

Having ink made commercially alleviated one set of problems but introduced another. Charles Manby Smith, in *The Working Man's Way in the World, Being the Autobiography of a Journeyman Printer* (a memoir that reads like a Dickens novel), remembers a buyer who "learned to judge of the qualities of printer's ink, before he gave a large order, by the flavour of Messrs. Lampblack and Bone's old port, a hamper of which was found by accident at his lodgings." A trade magazine printed a letter from a country ink maker who, declining to pay the kickback, had had his ink returned with "no end of sand and dirt mixed in," and in one cask "a dead cat, which we were bound in charity to suppose had tumbled in and not been put in—however, there was the cat." The next letter writer sneered, "Who is this lamb in the country … this 'dead cat

ink-maker' who wishes to rob poor men of their meagere perquis-
ites and Christmas presents?" One daily paper in New York paid
7¢ per pound for the eight hundred pounds of ink it used every
week. The same ink sold elsewhere for just over 3¢. The supplier
received $56 for the ink—and $24 was paid to the head pressman.
Bloy concludes that the purchase of ink was conducted at a low
level and that the master printer wasn't involved; he traces these
"nefarious practices" back to the wayzgoose, a dinner to which the
typefounder, joiner, ink maker, and others would be invited, and
would be expected to "add their Benevolence" to the workmen.

THE ACCOUNT OF ink making I'd read in Bloy, with the drunken
apprentices and the flaming pots of linseed oil, had sounded
terrific—"all the employees of the master printer departed outside
the city walls to make merry and ink"—but the too-neat parallelism
("make merry and ink") masked resentments that simmered along
with the ink. Bloy's source, Theodore Goebl, had said that such
ink making was like that wayzgoose, the traditional merry feast
for printers, but Charles Manby Smith's *The Working Man's Way in
the World* makes the wayzgoose sound fraught. The event started
around eleven a.m., with the men drifting in to play bowls in the
gardens of the tavern. By four p.m. when dinner was served there
would be two or three hundred men who'd been drinking steadily
for several hours. Speeches followed the dinner, with the junior
partner (the senior always absented himself so as not to restrain the
rowdiness) delivering a talk filled with platitudes about the nobility
of the working man: "It is in the order of Providence gentlemen,
yes—hm, he, haw, yes, gentlemen … that one man should wear
a better coat than another, though … it is the honest heart that
beats as often in the breast of a working-man as a king upon his
throne." Smith says these expressions of solidarity with the workers

are "nothing but moonshine, kept bottled up for the occasion, and regularly uncorked once a year." Then the overseer speaks, claiming "It has always been my object to stand between the employer and the working-man in the character of a fair arbitrator, to protect and to assert the interests of both."

From the back of the room the overseer's fatuous remarks would be "interlarded with the muttered thunder of discontent growled at a safe distance," because in fact the overseer exploited the workers and undermined the owner, often demanding bribes from the type-founder and the ink maker. The worker was as dispensable as the ink; what mattered wasn't the quality of the product but the favour of the overseer. The frenzied eating and drinking at the wayzgoose wasn't just celebration: the workers knew it might be their last chance for a long time. Fall was the slack season, and "at the heels of the weigh-goose … there not unfrequently comes the 'bullet,' as it is termed, or the sudden discharge, which sends a third or a half of the hands adrift after a fortnight's notice."

II

"Lunch?" said Steven. It wasn't a wayzgoose but a Texas barbecue nearly as extravagant—barbecued brisket, blackened chicken, breaded catfish, beans, potato salad, peach cobbler. I was hungry.

I said, "You know, I've read all about printing before, but the physical work of it didn't register."

"That's what we're trying to do for the students here. I tell anyone who asks about the program that I'm not turning out paper makers or bookbinders or printers. Terry Belanger at the Rare Book School says, 'We spend a week on what you do in an afternoon,' but the point is to give the students a taste of what it was like to be in the book trade, to give them an overview—Book Trade 101—and then they can pursue the separate aspects if they want to."

A gust of wind came in under the canopy that flipped his plate over and sent his can of Diet Coke spinning down the table. He calmly slid the food off the table back onto his plate and mopped up. "Lucky I always have one of these workshop rags," he said, unfazed. After lunch he showed the video of the "flashing" of the linseed oil; in it he and the other Steve, Steve Pratt, are similarly unfazed as the pot of boiling linseed oil repeatedly and unexpectedly bursts into flames. Even after it had been off the heat for ten minutes, when Steven skimmed the foam off the top, *PHOOMP!* the flames leapt again. "Look at that! That's excellent!" says Steve Pratt. "You let the oxygen back in and it ignites…. But I better get it off this wooden table." The video shows him taking it over to the safety of the sandbox—a sandbox full of dried leaves. "I cringe every time I see that," said Steven. "We probably could have set fire to the house. We just weren't thinking about it. Experienced ink makers wouldn't have done that."

"Yes, but if everybody was standing around toasting buns and drinking schnapps," I said, "you can see how things could get out of control pretty quickly. No wonder they managed to burn the print shops down. And occasionally a city."

"True. This workshop is only really possible because I'm the head of the library, and even so I expect a dean or a fire marshal to come in one day and shut us down. Heck, if I were supervising myself I probably wouldn't let me do this." The video showed them taking onions and charred toast out of the pot (they'd thrown in slices of white bread, the soft stuff I never see anymore except at Texas barbecues). I asked if they'd really have put that much bread and onion in the cauldrons of ink they made. I'd thought they would just drop in a little, that it was a sort of homeopathic proportion. "But you guys looked like you were making a stew," I said.

"Oh, I think it was that much," he said. "Steve Pratt has a list of the weights, but it was a lot." Again, my reading hadn't gotten me as close as I'd thought. I wanted to blow up some linseed oil. Clearly I had to visit this Steve Pratt.

I SPENT THE WINTER TEACHING, and reading (perhaps too much) about ink. I bought my own copy of Moxon, and in the introduction I read that when he and his father returned from Holland to London they set up shop in Bishopsgate (haunt of Shakespeare and the Queen's Men theatre troupe, now home to shimmering skyscrapers). I took the name as reassurance that I was on the right track. I ordered Hansard's two-volume *Typographia*, a beautiful reprint of the 1825 edition with green boards, gold-stamped spine, and creamy acid-free paper. The book had been reissued by the Kinokuniya Company in Tokyo and Thoemmes Press in Bristol, England. I'd never heard of either, and felt as if I'd encountered a secret international web of hand printers who with the internet and modern technology were keeping the centuries-old craft alive.

In volume II I discovered an advertisement for Thomas Martin & Co., an ink maker with a warehouse in London at No. 10 Fisher Street, Red Lion Square, presided over by a "MR T. BISHOP." My father had always disavowed his side of the family; maybe they were scurrilous ink makers. (Martin inveighs against those who by "the most despicable means"—hiring away his employees—have been stealing his trade secrets.) I'd once visited a book dealer in Red Lion Square, a ten-minute walk from the British Museum, the place where I'd stumbled upon Virginia Woolf's suicide note. I began to see connections in everything. An illustration of the same press I'd used in Texas made me think how the print shop collapses time.

You don't go back just to one era; Bloy, Moxon, Hansard, spanning three centuries, become your contemporaries.

My research took me to deeper obscurities, to Pedanius Dioscorides, an army physician in the first century CE, who in one of his ink recipes advised using the "urine of an uncorrupt boy." Cennini advises against grinding your red "lac" with urine, but in Matthew Skelton's novel *Endymion Spring* Gutenberg adds "just a splash," and indeed historians believe that one of the irreproducible elements in Gutenberg's fine ink was Gutenberg's urine. I now saw what other scholars had missed: if the printing press had produced blotchy copies, scholars would have gone back to the quill, but the ink was beautiful and people embraced the press. And so? The conclusion was inescapable—the Renaissance, the Reformation, and the Scientific Revolution derived from Gutenberg's pee.

Flashing the oil, a primary step in making printers' ink, occasionally set the print shop on fire. Photo courtesy of the author.

Flashing the Oil

I

A good motorcycle ride clears the winter cobwebs. In Texas, Steven Smith had told me, "Steve Pratt doesn't do email. You can write him a letter or maybe get him on the phone. He lives on a ranch near a hard-to-get-to place called Cove Fort. You'll have to fly into Salt Lake City, rent a car, and drive three hours south." I had a better idea. My previous book, *Riding with Rilke: Reflections on Motorcycles and Books*, had prompted an invitation to read at the Tynda Motorcycle Rally near Eugene, Oregon. In the spring I decided to cruise down to Eugene and then nip across to Utah. It was a hard ride: the wind in the Crowsnest Pass bobbled my head back and forth like a guy throwing pizza dough hand to hand. By the time I got to Coeur d'Alene, Idaho, I had grooves in my lower butt cheeks from the buckles on the saddlebag straps, and at the Visitors' Center I thought I was getting an electric shock from the counter, but it was just the tingling from gripping the handlebars. From Eugene it took me three days to cross the desert. I spent the first night in luxury at a Reno casino; the next night hardscrabble in Ely, reading on a hard kitchen chair wedged between the bed and the wall; and on the third fetched up in Beaver, at a faux-colonial Best Western, where the smiling Latter Day Saints manager adjusted the wireless connection for me, as if it would help me find God. This was a pilgrimage of sorts. I was here to flame out linseed oil, to stick my finger in the sludgy stuff that changed the history of printing.

Leaving Reno that first morning, I had begun my ride on Interstate 80, a broad boring freeway; it had once been a stagecoach route, and I wondered if my great-great-uncle Jeremiah had travelled that way. I wondered a great many things about him. After my ink-making session in Texas I'd visited my rancher cousin Janet up in British Columbia. She's the keeper of the family history, and though she thought my ink obsession was odd, she said, "I guess it runs in the family. You must get it from Uncle Jeremiah O'Leary."

I'd never heard of him. Janet got her box of papers, and handed me a small photograph of a man in a jacket and a white shirt and tie, with deep-set eyes, full moustache, and a bushy beard.

"Doesn't look a bit like anyone in the family," I said.

"He looks exactly like you," said Janet. "Here. In 1843 he was indentured to Crandall and Brigham, printers in Lockport in upstate New York, for a period of two years. Age eighteen. Then in 1852—so he'd be, what? twenty-seven—he takes the stagecoach out to Sacramento, to the California Gold Rush. His mother was really upset." Janet rummaged in the box and came up with three letters.

"Here's one from his brother-in-law. He says they were hoping for a meeting—*we was expecting a verbal one but it was in black and white* ... They would like to see him but *the land of gold is a long way from us ... prices is low here ... I would consider that board at*

Jeremiah O'Leary, printer and reporter for the Sacramento Bee.
Photo courtesy of Janet Cutler.

$50 a week high, though he thinks maybe he could go out with a
cargo of provender."

"You can tell he's kind of intrigued by Jeremiah's getaway,"
I said.

"Yes, but Jeremiah's sister Mary is not amused. She writes,
*What a disappointment to hear where you was ... we was agoing to
have some fine times with you*—she actually says *agoing*—*You deserve
a first rate scolding but it would not be much use by the time it got
there....* Then she decides to make him feel worse—*Jeremiah [his
nephew] is five years old today and told me he was sad you went away.*"

"Ah, playing the namesake-nephew card."

"Yes, but it's his mother Elizabeth who really gives him a blast.
Listen, *How you have cheated us all up here.... Oh it was too bad to
serve us such a caper to clear off to California....* She hopes she may see
him again, though she clearly doubts she ever will. Yet she admits,

We never gave you a broad or long credit mark for it as you asked...."

"So it looks like Jeremiah didn't just sneak off like a thief in the night. He'd told them of his plans but nobody really believed he would go. Did he ever come home?"

"I don't know, but he did all right for himself. Eight years later he bought a one-third interest in the *Sacramento Bee* for $3400. That would have been a lot of money in 1860. I wonder how he raised it?" We don't know. He worked twelve-hour days as a reporter, editor, and foreman of the composing room. He never married, and one day in June, just a month after his fiftieth birthday, he finished his shift, went back to his room, and died. The obituary calls him "unpretending and unassuming ... a just employer ... whose pleasure it was always to encourage the deserving." Does that mean he excoriated the undeserving? It's difficult to read between the lines of the cliché-filled obit. So many gaps.

In 1852 Uncle O'Leary was part of a crowd of 200,000 that had been pouring in to California from all over the world; in two years San Francisco had gone from a thousand-person village to a city of 25,000. To reach Sacramento from New York state by stage would have taken three jostling weeks, perched on a bench seat, your knees dovetailed with the other passengers if you were in the rear-facing front row, swaying with only a leather strap for backrest if you were in the middle row, forced to get out and push up the hills or wield a long pole to get the stage through mud holes, exhorted not to drink but to share the bottle if you did, not to spit tobacco into the wind, not to cuss in front of ladies, not to snore on your neighbour's shoulder, not to hog the buffalo robes or shoot your guns out the window (it spooked the horses). His ride made my discomforts on a motorcycle look mild.

The first stagecoach robbery in gold country took place just before Jeremiah passed through the area; I'm sure he heard of it—$7000 in gold bullion—but he didn't mention it in his letter home. Probably didn't want to spook the ladies. Did he ever bring out his sister and the nephew named for him? Little Jeremiah would have been thirteen when his uncle bought into the *Bee*, and I like to think of the nephew—my great-grandfather—taking the train out to Sacramento, touring the press, then rambling on his own in the city, watching the steamboats come up the river, checking out the California girls, maybe buying a pair of Levi Strauss's new denim pants with copper rivets on the pockets. But the Gold Rush was long over by 1860 and the times were as unsettled for O'Leary as they had been for Joseph Moxon: while O'Leary was finalizing his deal with the newspaper the Southern Democrats were splitting from their party, and in November Republican Abraham Lincoln was elected on an anti-slavery platform. Six months later, a year almost to the day after Jeremiah O'Leary bought his share in the *Bee*, Confederate forces attacked Fort Sumter and the country disintegrated into the Civil War. Not a good time to travel. My guess is he did not see his family again.

What would his experience as an apprentice have been like? His master David Crandall was, according to one source, "one of the jolliest and most jovial men that ever lived in Lockport," but history is not written by apprentices. Historian Robert Darnton says the apprentices slept in a freezing room, ate slops from the master's table, and suffered constant abuse. In *The Great Cat Massacre* he tells the famous story of how some French apprentices rounded up stray cats, staged a mock trial, and hung them one by one in the courtyard of the print shop, including (though they hid the evidence) the favourite cat of the master's wife. Darnton explains that *le chat, la chatte,* and *le minet* mean the same thing in French slang as *pussy* does in English. Thus in killing the wife's cat they were symbolically violating her.

Uncle Jeremiah had a kindly face and I couldn't believe he would abuse cats, but as an apprentice he probably had his own revenge fantasies about the bosses. Then of course he became one. As foreman of the composing room, Jeremiah would have had to fire some drunken typesetters. I hoped that phrase about him being one "whose pleasure it was always to encourage the deserving" didn't hint at another side, and that he was one of the fair overseers.

Conditions in the print shop were harsh. Manby Smith says that in the winter, "we had sometimes to thaw our frozen type by burning paper upon the face of it" before distributing it back into the type cases. He worked in Hansard's shop, and he describes how they worked round the clock to produce the Blue Books for Parliament—a fifty-hour stretch without sleep in the "Babylonish din," with the "banging of the mallets, the sawing of 'furniture,' the creaking of the old press, the shuffling feet of the messengers, the bawling of twenty voices, and the endless gabble of reading-boys in the little closets which abut upon the composing-rooms...." The atmosphere "would poison a vulture," for though it is cold and raw outside, "the perspiration streams from every face within." Hansard was a no-nonsense sort who didn't care about fancy coloured inks: "black, as perfect as blackness can be, is, in my judgement, the true criterion of good ink." In his *Typographia* he swoons over the "beautiful velvet richness of colour," the "harmony of tint and richness" in the ink of Bulmer's edition of Shakespeare. At Steve Pratt's I would encounter such velvet richness.

II

At nine a.m. I rolled down the gravel driveway, past the little sign on the mailbox that said *Pratt*. Steve wore jeans and a black cotton shirt—not macho black but soft, as if it had been worn and worked in for years—and a blue baseball cap with wagon wheels on the front, the emblem of Pratt Wagon Works. He was tall and skinny with an infectious smile and a Jimmy Stewart voice. He described himself as a "master craftsman" without arrogance, as if that were a grade, a designation, not a self-applauding adjective.

His son Ben, wiry like his dad, wore dirtier jeans and a plaid shirt and a cap that said NAPA. Midway through the morning he jumped on his dirt bike and rode down the road. Steve explained that he made extra money checking on a nearby gas plant several times a day. "He gets a dollar a minute, which is more than I pay him," he laughed. "He's a fine craftsman." Ben was repairing a .22 Winchester from the 1880s and showed me how he'd replaced the bore. He worked carefully with the file, sitting on a stool, and I was reminded of the Eastman guitar factory in Beijing, of how what you didn't hear was the whine of power tools. Steve was obviously proud of his son and of his skills.

I told Steve I had ground sheep bones last year in Texas, working away for an hour with the big iron grinder thing. He couldn't remember the name either, and then we found it in the notes. Of course: the muller. He'd spelled it mauler, which seemed a good name. "Those were my sheep bones," he said. "You carbonize them on the barbecue. The important thing is to keep the lid on to keep oxygen from getting into the bones. You want to cook out all the interior matter, but if you have air in there you'll burn up the bones entirely. Yes, they are hard to grind. I use this now." He opened up a can marked *1 Lb. Dry Pigment, Bone Black*, put his finger into the pigment, and put it on his tongue. "This is how you can tell if it's ready." I paused, but thought, Nope, this is what I'm here for, and stuck my finger in too. "See? What do you feel on the tongue?"

"Nothing."

"That's it—no grit. It's perfectly smooth. That will make great ink. It would take hours to get it that fine grinding by hand."

The really excellent black resembled velvet, both in colour and texture; Bloy's recipe #9, from seventeenth-century France, declares it was "like the finest chalke, or flower [flour]." And you should check it carefully, "for there is a counterfeit sort made of Lees of wine burnt, which is nothing so faire, but harsh and injurious to the [printing] plates." I figured the stuff we'd made in Texas with my ill-ground bones would have been injurious to the plates indeed.

Steve pulled out some jars of ink he'd made before. "I don't know if these will still be good. They've been sitting since, let's see, 2002." He stuck his finger into one. "I put water on top like they tell you to do to preserve it. But no. See? It's hard. This is no good anymore. Let's look at this one." He stuck his finger under the water into the black goop. "Ah! Look at that. Rich, creamy, this would make perfect impressions." I picked up the little bottle and put my finger in too, like reaching down through lake water into mud. The ink was cool and, yes, creamy. I pulled out my finger and rubbed the ink between my finger and thumb. Absolutely smooth. No grit, unlike my ink in Texas, which had reminded me of when I'd tried to make a cake and added sugar to the butter and hadn't stirred it enough to make it smooth.

Steve walked me back outside to get my camera and notebook out of the saddlebags. "So what kind of bike is that, Italian?" He said he'd never heard of Ducati. But he had owned motorcycles. "I started out with an old Indian in high school—*that* taught me a lot about fixing things—and I ended with a Harley. For seven years I commuted into Salt Lake on that." He sympathized with my story of riding the freeway, chock with hurtling SUVs even early on a Sunday morning. "In between I had, oh, a big Triumph, a Kawasaki—I can't remember the numbers now—and a BMW."

I asked which one. "It was the biggest one they made back then, a 750. I rode that one up to Canada, to southern Alberta. I thought at that point that I might go live in Canada and wanted to go see what it looked like."

I didn't ask what the story was. A potential draft dodger? Or maybe he was active in the church; lots of Mormons settled in southern Alberta. His mother was Canadian and he has relatives near Calgary. He let me change out of my leathers in his office, and above his desk he'd pasted a series of opening pages of magazine articles, each with a picture. I noticed first Chiang Kai-shek and read the opening lines about his fight against communism, then General Pinochet of Chile, also lauded for his fight against communism, then Senator McCarthy, who'd been unjustly maligned and was now seen as heroic for his fight against communism. I stopped reading. I was here for ink, not politics; I ignored the big poster on the door about LIBERTY and the Constitution. I already liked Steve Pratt, and I remembered an article by a congressional intern in which she said she'd expected to like the people whose politics she agreed with but had discovered that often those were the people she found insufferable, and that the people she enjoyed were those whose politics she abhorred.

STEVE WAS IN CHARGE, but Ben had journeyman status and worked on his own. They had an affectionate banter.

"What propane tank should I use, Ben?"

"One with propane in it, Dad."

As we looked for a full tank, Steve talked about *abgekröschte*, finding the word in a recipe and wondering what it was, learning that it was related to *abkreischen*, which meant "to cry out," and *abkröschen*, which meant "to roast," and finally discovering that it was the rolls that were cooked in the linseed oil. Steve had bread and onions out on the workbench. "You can use any kind of bread, preferably stale. In the old days they said that it absorbed impurities in the oil. Today's oil is so pure that the bread has no effect on that. What we did discover is that the oil won't boil without it, not a real rolling boil like water, and what we figured was it's the moisture in the bread and onions that enables the ink to boil."

Steve told me he'd produced three lines replicating Gutenberg's type for the print museum in Provo, Utah. "I wrote an article about it for the Printing Historical Society in England, and they were very glad to have it. They said, 'This will teach those saucy Princeton professors a thing or two.' I believe Scholarship Should Be Married to Craftsmanship." You could see the capital letters by the way he said it. I wondered if he knew he was echoing Moxon's dictum that "bare reading" wasn't enough.

"Then I did these experiments with ink. We used the recipes in Bloy, but the fact is you can't make good ink from any of the recipes in Bloy. Not one." I was cast down. I thought Bloy was our hero—mentioning his name as I unpacked my motorcycle had seemed to cinch things with Pratt, make me legitimate; it was the secret handshake. Bloy was the guy who had done the work, and now I was hearing that his recipes were no good. Who could you trust? But Steve was still talking.

"… you have to look at all of them, and then you see some things are mentioned in one that aren't mentioned in others. So we went through and put together a composite recipe. And *that* makes good ink. Look at this, still rich and creamy seven years later."

I didn't stick my finger in this time. The ink wipes off your finger easily with a paper towel, but it gets under the nail, and I suspected (rightly) that it would take a week or more to get out, until the natural skin oils finally lift it off.

"We wrote this up and sent it to the same people in England, and you know what? They never wrote back! No one is interested in ink. I can't understand that because it's so important." He would find a kindred spirit in Philip Ball, who in *Bright Earth: Art and the Invention of Color* argues at the outset that "This neglect of the material aspect of the artist's craft is perhaps a consequence of a cultural tendency in the West to separate inspiration from substance." Ball quotes Anthea Callen, who, in writing about the Impressionist painters, insists that "Any work of art is determined first and foremost by the materials available to the artist, and by the artist's ability to manipulate those materials."

Inks had been the software of their day, the secrets jealously guarded—you'd get together to boil down the linseed oil and eat *abgekröschte* rolls, but you'd never tell anyone about your secret additives. I wondered if that was why all the recipes in Bloy were missing something. When ink making became commercial, manufacturers were constantly updating, offering new applications. Mitchell reports that in 1811 there were eleven patents given for inks—including one for a flexible ink and one for writing on glass—and that from this point on, ink patents proliferated: in 1862 there was one for a telegraph ink; in 1873 a fireproof ink (asbestos powder); in 1897 an ink both invisible and indelible (alum and white garlic juice—visible on heating); in 1898 an x-ray-proof ink; in 1905 an ink for printing on leather; in 1909 a mimeograph ink; in 1910 an ink for printing on bread; in 1912 an ink for coating carbon papers. There was "web-press ink," thin and able to penetrate the paper rapidly for continuous runs like newspapers; "flat-bed ink," which had a heavier body and was used for the harder paper of pamphlets and books; and inks for every class of work to be done, varied for the grade of paper, the speed at which the presses were to be run, and the temperature and humidity at the time of printing. Every one a secret. Yet they all started with some kind of varnish.

III

We had three apprentices for our work today: Levi, nine; Nathan, seven; and Raina, four. Levi was solemn in a cowboy hat and plaid shirt with plastic mother-of-pearl buttons and a T-shirt underneath like Ben; Nathan more the kid, with carrot-coloured hair, blue T-shirt, and new blue-accented cowboy boots and a pair of rawhide gloves: his birthday presents; and Raina, shy but eager, with a pink plastic bow pinned in her hair and a pink T-shirt over her blue jeans. "My grandfather's pretty amazing," said Nathan when Steve went to take a phone call. They loved their grandfather, and they knew how to move in the shop. They weren't rigid but they made no sudden movements, stayed close to the workbench, didn't pick up tools and wave them around. I wondered how much Steve had said to them and how much they had just learned from being around him. Levi was the age Moxon had been when he was working in his father's print shop.

I asked who the best builder was of the kids and Nathan said, "Oh, Levi," and showed me the toy boats they'd been building. But he was proud of his own work too. He was excited about making ink—"It's going to explode! *Psshhhhh!*"—and he threw his arms out. Their grandfather had obviously told them that they'd be setting stuff on fire. That was the attraction. So cute, I thought, and then realized that that was why I was here, too. To watch stuff explode, to catch on fire all by itself. Nathan didn't know the term *spontaneous combustion*, but we knew it would be cool.

"In the beginning stages it's basically just deep frying, like at McDonald's," Steve said as we put the slices of bread and the chunks of onion that Levi had cut up into the hot oil. "The recipe calls for two percent dryer, but they wouldn't have had that in the old days, and they didn't need it. The onion would have produced acid, and that acts as a dryer." As with the ballpoint pen, the crucial thing is to have an ink that goes on easily but doesn't stay all smeary on the page. I learned later from *Henley's Twentieth Century Book of Formulas, Processes and Trade Secrets* (1934) that the heating is to initiate a first oxidation of the oil so that it will dry better. Linseed oil is itself a drying oil, and when exposed to air in thin coats it will absorb large quantities of oxygen and thereby be converted into tough, solid sheets … very similar to those of soft India rubber. The shorter time you spent heating the oil beforehand, the longer the ink would take to dry.

We were outside now, in the lee of the shed, important because there was a gusty wind. "Ink making has burned down print shops, and even whole cities," Steve said. Nathan wanted an explosion, but even he wouldn't want to see the workshop go up in flames. Point number one on Steve's recipe from Bloy begins, "Outside on a calm day, away from any fire hazards, heat pure raw linseed oil…." Steve said, "I always like that—'outside on a calm day.' The first time I tried it I did it in the shop; it blew the lid off and I decided we'd better do it outside. Every time you do it it's different. When I made ink with Steve Smith down in Texas we got the big flames and he said, 'I thought you said you'd done this before!' and I said, 'I have, I have!'"

To flash or not to flash, that was the question. Some recipes insist that you *must* flash the oil, that without that final step you'll never achieve the dense consistency needed for a good varnish; other recipes insist that you must *never* flash the oil. Who was right? Steve was stirring the bread and onions in the pot, lifting them with the ladle, noting how the edges were starting to blacken, when he said, "You know, I've found it doesn't make any difference at all whether you flash the linseed oil or not. We've tried it both ways and the varnish it produces is just the same."

Maybe this was like those recipes where one chef will insist you must do *X* or the whole thing will be a total failure, and across town a rival will declare, Do *X* and you will ruin the dish irrevocably. I only knew that when I used to make coq au vin—before I read a recipe that told me to warm the brandy and light it *in the measuring cup*—I'd sometimes pour heaps of the stuff on and it would sink in before I had a chance to get the match going. It wouldn't light, so I'd pour more, and it still wouldn't light, and then finally I'd get it going, but it would be a feeble flame and the coq au vin would have a slightly soggy, cheap-brandy taste, even though I'd used good Courvoisier—another mistake; I learned that the cheap stuff flames much better.

Then I learned to use a barbecue starter to ignite the brandy. No more fiddling with paper matches and trying not to burn my fingers when the pan went up. I got the brandy ready, lots of it; I got the barbecue starter ready, in my right hand; I poured the measuring cup of brandy in a big circular swoosh with my left hand, clicked with my right, a moment of stillness and then *WHOOMP!* I had a column of flame the circumference of the frying pan, as if the pan were the back of a jet engine and my stove was trying to fly into the earth. I watched helplessly as the column seared the white cabinets above the stove a sooty black. It did not occur to me to cover the flame. I wonder how many people watch immobile in the first moments as they inadvertently burn their houses down. I passed the mirror in the hallway and saw that I'd singed my eyebrows as well. I don't remember how the coq au vin tasted, but I'll never forget the towering inferno. I understood Nathan's excitement. I did get the taste right when I mastered the art of the discreet *flambé*, but I still always feel something is missing. There's nothing like setting fire to stuff.

As we waited I asked about the beautiful unvarnished wagon wheels inside the workshop by the door. "Not just wheels. I'm a master craftsman and we make wagons. I once taught electronics, but finally decided that writing equations on the board was no fun. So now I do this." People come to him from all over the United States and England, tracking him down even though he doesn't use email, has no presence on the web. "Someone called from Britain about repairing a press and I said, 'Why are you calling me?' and they said, 'Because no one else does this.' So that's what we do here, the things no one else does. See that wagon over there? It's supposed to be ready for a museum in Germany next week." Clearly it had not been touched in a century. "That's why I was a little uneasy about taking time off to make ink with you." He'd been effusive during my first phone call and turned cool on my second, said he had paid work to do, and so I offered to pay. I'd wondered at the change. And now a woman, grey-haired and grim-faced, opened the door to the shop. "Richard is on the phone and wants to talk to you right now."

I walked toward her and put out my hand. "Hi, I'm the Canadian, I talked to you yesterday…." She looked past me at her husband. "*Right* now." I held my hand out a moment longer, then dropped it to my side. She closed the door without looking at me. I was shocked; I thought people were only cut like that in Edith Wharton novels. Yet I understood. I was the troublemaker. People like Steve are so enthusiastic about what they do that they can be too generous with their time and talents. So one partner takes over the business side ("She's buffering for us now," he said with a smile), and that allows the non-business partner to become even more unbusinesslike. During that first call, when I first mentioned grinding the sheep bones, pouring the varnish, working the muller in Texas, and now wanting to experience firsthand the flaming of the ink, Steve was almost capering on the other end of the line. I'm sure he would have taken the whole day, maybe two whole days, to experiment with ink, to try it on the press—and all the while the wagon from Germany would be sitting outside the shed, the house he was supposed to be building for his wife ("She had a bad fall a while ago and now feels the big house with all its stairs is dangerous") sitting unbuilt. So when this unpaying customer comes out of the blue and expects the master craftsman to drop everything and "have fun" (as he put it), of course she would— should!—be annoyed.

And yet. There's also the envy the manager partner feels for the artist partner, the one with a vocation instead of an occupation, who would do what they do for free, would be doing it anyway, irregardlesslessly (as a painter friend put it), for the delight of figuring out something new, for the craft. Schedules and salaries don't enter into it. Meanwhile the manager partner watches as if through glass because it's something they've never felt for themselves, can't comprehend, don't even quite believe in, and while they're proud of the product, pleased by the adulation of the world, in private they gnash their teeth because, indispensable as they make themselves, they know they are, ultimately, emotionally dispensable. The vocation partner loves them, oh indeed, probably more intensely than the manager partner loves—but if the manager partner died, the other would go on, for in their work they achieve a pure happiness, oblivious of the other. And so the manager partner snarls and gnashes his or her teeth and hurls herself into protecting the vocation, making it their vocation, facilitating but finding ways to sour it a little, slipping a tincture of bile into the mix, never once feeling fully appreciated, never once knowing joy. But hey, what do I know? Hsing says I'm always making up stories about people, and in fact my contact with Mrs. Pratt had been maybe seventeen seconds.

IV

"Okay, look at your watch," said Steve. "We're, what? Say forty-five minutes to the point when the onion and bread start to turn black." The smell of the onions in oil made us hungry. That would not last long. The smell comes in three stages: enticing, repellent, and gagging. "Okay, now they're all black, see? In a minute we'll put them over here in a bucket—carbonized onions and bread are a fire hazard, believe it or not. Three times I've set them aside, with no heat source under them, and they've burst into flames." I was thinking how I still loved the smell, the hot oil and onions and bread, even though they were too charred to eat, but even as I was thinking it the pot started to smoke and the smell turned harsh, all food flavour gone.

"Okay, I'll ladle these out. What is it? Call it an hour to this point. Levi, can you get us a bucket? We do home schooling here in Utah. It works out good. Levi is the second generation to be home schooled. Ben was the first, and now Levi." Now the smoke was more than unpleasant, it was making us choke. It felt like it was coating my tongue and throat and lungs, thick goop gagging me. Levi, Nathan, and I dodged out of the shifting stream of smoke. If you were stuck inside there'd be no escape. I wondered how many apprentices died young. *Henley's* says the large quantities of vapour are mostly acrolein, which was used as a weapon in the First World War. We were doing as the old recipes in Bloy advised: "He boiled his oil in a meadow to keep the evil odour away from the community, and to avoid burning down his establishment."

Levi got some small bottles and a roll of cheesecloth. Steve cut the cheesecloth into large squares and fitted them into the mouths of bottles to form funnels. "They say the cheesecloth is necessary to catch impurities, but I doubt we'll have any. This oil is pure to start with, much purer than they would have had." We were using "Sunnyside pure raw linseed oil" made in California. "I wasn't sure we still had any, but Ben said he remembered seeing some on a shelf and he was right." Even for a small thing like remembering the linseed oil, there was pride in Steve's voice. I wondered if it was defensive pride, if he ever wondered whether he'd done the right thing by keeping Ben here, home schooling him, tutoring him in the crafts of outworn ages. Levi was Ben's son. The mother must have been fine-featured and beautiful. I wondered where she was, what the story was, but it was none of my business.

Steve poured off some of the warm varnish through the cloth into one of the bottles. "This is the first pour," he said. "We'll do three if we have enough. I usually make enough for a quart of ink, but you don't want to carry a bottle that big on your motorcycle." It looked like liquid toffee, and I thought of the sugaring-off of maple syrup, where you start with maple sap as light and clear as water and you boil it down and down and down until it gets dark and thick like this, and then you pour it on snow and get a wonderful toffee that bolts your teeth together.

"The old ink kettles had small openings at the top so you could cut off the oxygen quickly," he said as the lid rattled on the pot and flames shot out to both sides.

(... *clapping the cover upon the pot it will be extinguished, provided it be very close* ...)

He held the heavy tongs in thick-gloved hands, forcing the

lid down. "Here, it doesn't want to … there." It was out. "Now let's take it off and see if it spontaneously combusts." Nathan was fascinated, pressed against the wall of the shed but eyes glued to the pot. Steve took off the lid. We waited for the *WHOOMP*. Nothing happened.

"Levi, get your torch. We'll start it again. There, put it under the kettle, remember we shut the heat off."

We hadn't remembered. In the heat of the moment Steve had turned off the burner and now the linseed oil was too cold. Levi got it going again with his torch and then ran off to stomp it out on the gravel driveway.

"We won't need the torch again; we'll do it with self-ignition."

(… *it may take fire gently of it selfe, or be easily inflamed with the blaze of paper, as wine is burnt* …)

It didn't work.

Steve heated the pot again but the oil wouldn't ignite. We were all disappointed. I turned away to write in my notebook.

"Whoa-hey! Look at this!"

A pillar of flame the full circumference of the pot shot straight up, then the wind snapped it flat, left and right, and it split into two flaming wings, still for a moment like an orange Winged Victory of Samothrace before the wings flapped and stretched into fingers darting this way and that and then swooping back into a column to billow up again. Steve clapped the lid on the pot but wraithlike fire slithered out. He jiggled the lid, caught the lip, and finally the flames vanished. All this in a few seconds from an inch of oil. No wonder print shops burned.

"You see?" Steve held up goop on a stick. "There was a thick covering on the oil, so the oxygen couldn't get in. As soon as I stirred it, it took off!" Triumphant. This was cool. Raina was nowhere around. Is blowing stuff up gender specific? Orange flames, subdued now, danced across shiny black tar, the bubbles now solidified into a pebble grain as if marbles were buried underneath.

"The self-ignition is so unpredictable. You can see why they said 'outside' and 'on a calm day,'" Steve said.

(… *least it endanger the house … your eye must be continually upon it …*)

He poured some into another jar.

(… *when cool enough to place one's hand in, strain through a linen cloth …*)

"Uh-oh." The bottom of the jar looked like it had been severed with a laser, neatly cut off at the seam. The brown varnish was oozing across the paper. "This is why we put a covering down on the table," he said to Nathan and Levi, "and this is why we should be careful in pouring hot varnish into jars."

(… *let it coole a little, before you poure it into the vessell, in which you intend to keepe it …*)

Steve reached down and drew some varnish together between his thumb and forefinger, then held his hand up in front of his face and carefully drew thumb and forefinger apart, seeing how far the drop would stretch before it broke. "This is how we test for the 'thread.' Bloy says the best thread is three-eighths of an inch." Many of the recipes that speak of threading say things like "the oil is ready when it threads"—but they don't tell you what threading is. I took some too. If you get a long stretch like toffee, then the varnish is still too thin. Steve said, "One time we got a thread ten feet long." If you have less, then you've boiled it down too much and it's too hard.

Steve picked up the pot again. "Oh no. We've overcooked it." We were the bad apprentices, chatting by the fire instead of paying attention. "This is such a small batch that there wasn't enough liquid." It was terrible. There would be no threading this. Steve

Steve Pratt testing the thread: shorter than toffee, longer than glue—just right.
Photo courtesy of the author.

poked around in the bottom of the pot with a stick. There was a lustrous black-bronze sheen where he stirred, but it was thick like the glop in *Alien*, and over it, shiny and black, was a thick rubbery film, like old pudding that has been left too long in the fridge uncovered. The film came apart as he moved the stick, contaminating the little layer of varnish underneath. There was no way to save it. My camera captured a grinning demonic face in the bottom of the pot. An ink imp.

I wondered about saving the pot, but before I spoke Steve said, "This is easy to clean. We just turn the heat on again and it will burn right off." At least we wouldn't have to scrub it.

So that was it. We were now at the point where I'd entered the process last year, grinding the pigment to be added. Steve found a smaller bottle for the little bit of varnish he'd managed to save from the first pour, and he gave me one of the bottles of ink he'd made seven years before. The label on the bottle says *Pratt Wagon & Press Works, Cove Fort UT* with an etching of a covered wagon and a printing press as tall as the wagon.

We were done.

V

Steve took me into his print shop to show me samples of the work they did with the different kinds of ink, and to show me one of the presses he'd made. "So you make everything, from the ink, to the type, right up to the press itself," I said.

"Once we were asked to set up a whole print shop. The clients had rebuilt it and saved the original flooring, good quarter-sawn oak, beautiful stuff. We put the floor back and then they said, 'We want you to place the press where it was, from the marks on the floor.' And you know, you could see where the pressman's heel was, where he'd pivoted a thousand times a day. It was worn down into the oak and all polished." I loved this, reconstructing a print shop through the motions of the long-lost body. An archive in a plank.

What caught me though were the lovely shoulders of the Albion press over in the corner. It looked almost delicate in its pastel-green paint ("That was the colour of the original"). I turned the crank that slid the bed and pulled on the handle; both moved with a willing smoothness. Soft, sensuous. "We make them right here. That is, we send out to have the castings done at a foundry near here and then Ben and I assemble them here." I ran a finger along the wooden handle. It was curved and smooth as a lute. "We make sure everything works before we send it out." It had cute little feet with flutes that looked almost like toes. "We put our name on it too." One side said *D & J Grieg, Edinburgh*, and the other *S & B Pratt, Utah*, "because it's considered bad form to make a reproduction and not clearly identify it." The headstock was engraved with scrollwork. "That costs a couple of hundred dollars extra," Steve said. "Or you can get it just plain."

I could get it, or part of it, using my university equipment allowance. But then it would be partly the university's. I knew that when I died I'd leave it to the university (Hsing's brother works for the Development Office, which means he phones people up and asks them whether they're planning on dying soon and if so would they like to donate to the university, so I'm mindful of such things), but still I wanted it to be mine. I hadn't felt such techno-lust since I'd bought my motorcycle, and this time I knew it was pointless to conceal my infatuation. "I think I'm going to have to have one of these," I said. "We're renovating our basement and there'll be room for a press." Hsing wanted room for her string quartet, but this wouldn't take up much space. Maybe I would print them a sheet of period music. They might want to do a little printing on their break.

I daydreamed all the way to Tooele, a little city surrounded by green ridges sloping down toward the Great Salt Lake. As the evening light changed on the hills, the colour reminded me of the Albion, of how scholarship must be married to craftsmanship, and I wondered what Jeremiah O'Leary dreamed as his stagecoach passed this way.

PART II

THE *Art* OF

The weakest ink is mightier than the strongest memory.

—CHINESE PROVERB

Inksticks with four-toed dragons—five-toed dragons were reserved for the emperor. Lao Hu Kaiwen ink factory, Shexian, China. Photo courtesy of the author.

The Inksticks of Anhui

I

Chinese ink is solid, Western ink is liquid. From this all else flows. Inksticks—picture a domino with gold Chinese characters on one side and an image of a person or a landscape on the other—along with inkstones, brushes, and paper, are one of the so-called "Four Treasures of the Study." No Bics and yellow Post-its here. Ink was kept in solid form and ground fresh every morning. You would place a little water on the inkstone, which had a lip to contain the ink, and begin grinding your inkstick in a circular motion. The soot and glue of the inkstick combine with the water to form ink, which then runs into a small hollow at one end of the inkstone. A Chinese scholar would grind his (or in rare cases, her) ink fresh before each session because it spoiled if left overnight. The hot Chinese summers affected the gum. More importantly, though, the grinding—at least one hundred rounds—was a meditative act; it focused the mind, prepared you for the moment of writing. "Good ink cannot be the quick kind, ready to pour out of a bottle," says the fierce grandmother in Amy Tan's *The Bonesetter's Daughter*. Bottled ink is more than just a lazy shortcut (though it is that—"You can never be an artist if your work comes without effort"); the real problem is that it undercuts the creative process: "You do not have to think. You simply write what is swimming on top of your brain. And the top is nothing but pond scum, dead leaves, and mosquito spawn" (this in an era before Facebook updates). But, she insists, "When you push an inkstick along an inkstone, you take the first step to cleansing your mind and your heart."

The first time I ground ink it reminded me of grating carrots. My Taiwanese aunt-in-law Lisa directed me to grind in a clockwise motion. There was pleasant resistance and then the slide of the mush underneath. I read once that ink should grind on the inkstone like wax on a hot metal plate. Mine felt scratchy. Bits broke off. I did not think I was cleansing my mind or making fine ink. Later I would learn how ink and stone work together. A good inkstone is important because the form of the stone's crystals can affect the quality of the ink produced. Conversely, inksticks with impurities can damage the "tooth" of the inkstone, and the impurities will mean grains are left on the paper and thus the ink will not achieve real blackness. (This seemed awfully fussy until I thought of coffee. Good beans are essential, but the burr grinder I'd been given for Christmas gave me a better espresso than my old blade grinder.) If the ink is ground too thick it will not run freely, too thin and it will blot and expand into lines thicker than intended.

Ultimately, lustre is what you look for, that and permanence. The best inks last for thousands of years, and retain their deep hue. We respond to deep, lustrous ink; this seems to be something that transcends cultures. It's what makes us love the Gutenberg Bible. *Lustrous:* the quality of shining by reflected light, from the Latin *lustrare*—illuminate. A better word in feel and meaning than *glossy* or *gleaming*, which have the association of something superficial, meretricious. When seeing a sheet pulled off the printing press for the first time, students in my Book History class spontaneously "Ooh." Then, when asked to compare the printed sheet with that made by a state-of-the-art photocopier, they immediately point to the printed page, usually with a snort of derision for the photocopy. It's visceral.

The grandmother in Amy Tan's novel says the best ink is a purple black, neither brown nor grey. My ink was translucent grey, like the mud puddles I'd used as a kid to write on cardboard with a stick. "It's okay," said Lisa, by which she meant that with the scratchy brush and my beginner's technique we would not be concerned with lustre or longevity. The strokes of a Chinese character are formed precisely, and laid down in strict sequence, and there is no going back to touch up. In Miss Bradshaw's penmanship class we were always going back to touch up a *t* with a split stem or adjust the curlicue on a capital *O*. Not in China. "No," said Lisa as I tried to fix a stroke. It didn't work anyway, just made a larger smudge.

Traditionally, credit for the invention of ink goes to the third-century calligrapher Wei Tan. In his recipe, after you've strained your soot and dissolved your glue in the juice of the *chin* tree, you "add five egg whites, one ounce of crushed pearl, and the same amount of must, after they have been separately treated and well strained." Then there was a range of additives to improve the colour, consistency, and aroma—everything from peony rind, pig, or carp galls to pearls, pomegranate, and sandalwood. Over eleven hundred possible additives, and every ink maker had his own secret blend. "All these ingredients are mixed in an iron mortar; a paste, preferably dry rather than damp, is obtained after pounding thirty thousand times...." (*Thirty thousand?*) "Or pounding more for better quality." (*More??*) I bet Wei Tan wasn't doing the pounding. I would encounter this in other cultures—the endless, soulless grinding that the apprentices had to do. Also, just as the Italian artist Cennino Cennini does twelve centuries later in his handbook, Wei Tan attends to the seasons: "The best time for mixing ink is before the second and after the ninth month in a year. It will decay and produce a bad odour if the weather is too warm, or will be hard to dry and melt if too cold, which causes breakage when exposed to air."

But Wei Tan did not really invent ink. The famous oracle bones have red from cinnabar and black from some kind of carbon ink or dried blood.

Even before the ancient *I Ching* (the oracular *Book of Changes*, popular still with Western undergraduates seeking their fate), divination was practised by oracle bones: the shell of a tortoise was inscribed in vermilion or black ink and then pits were drilled into the bone. At the moment of divination a hot stick would be placed in the pits, the shell would crack, and the oracle would be read according to the way the cracks separated the characters. These go back far beyond Wei Tan: the bones in the Royal Ontario Museum date from around 1250 BCE. Apparently the earliest form of the character for ink (*mo*) refers to a punishment of blackening or tattooing the face rather than to the use of a writing fluid. The word combined two characters, for *black* and *earth*, and the original seal form depicted a flame under a smoke vent or window, blackening it with soot—an image still represented in the more stylized modern character. With the foot added to the bottom it looks like an angry robot:

墨

In my first efforts with calligraphy, the ink and I became angry with each other. The horizontal stroke, *heng*, is the first of the "Seven Mysteries" of Chinese calligraphy. It is supposed "to seem like a formation of cloud stretching from a thousand miles away and abruptly terminating." It has five separate movements. I could discern only three. I could execute only one.

I held the brush upright ("No slant"), two fingers in front, two fingers behind. I dipped the brush, lifted it, lowered it, drew it across the page in a single short stroke.

"Hmm," said Lisa.

"Not so good?"

"Not good. You try again." I tried again. She showed me again. Lisa's hand trembled as it approached the page and looked sure to produce a wobbly blot. Instead the character emerged swift and crisp. Her stroke was rounded at one end, sliced off at the other, and narrowed slightly in the middle, like a bone.

I tried again. And again. My line looked like a row of sausages.

"Better?" I said.

"Hmm," said Lisa. Did she not know that in North America we believe in false encouragement, praise for effort?

The grandmother in *The Bonesetter's Daughter* says, "When you grind ink against stone you change its character from ungiving to giving, from a single hard form to many flowing forms. But once you put the ink to paper, it becomes unforgiving again." I wasn't flowing and Lisa was unforgiving. She moved around behind me, reached over and gripped my hand. Not gently. "Relax," she said. I tried to let go but my arm stiffened against the grip. I thought we were going to the paper, but she was moving my brush over to the ink. Although I was trying to be helpful, it was as if we were arm wrestling.

"First down," said Lisa, and before I could pull down, which is what I thought she meant, she pushed the brush down onto the page. We made a blob. "Up." She lifted my hand slightly and we swept across, the wobbly line a reflection of our inadvertent tug-of-war. "Back." Her hand guided mine under and back over our line. I tried to go around above the line; she pulled me below. We ended in another blob. "Again," she said. "First you push to, to …"

"Anchor the stroke?"

"Yes."

By the third time we were no longer arm wrestling. The initial mark was more rounded, looked less like someone had squashed a bug there, and somehow we made a crisp ending to the stroke. She let me do it on my own. "Better," she said. Maybe. Instead of sausages I was producing dumbbells. I couldn't do it on my own. "You must practise." Yeah, I know, ten thousand strokes. Everything in China is ten thousand of something. Long before Malcolm Gladwell touted ten thousand hours as the key to genius, the Middle Kingdom had decreed it the minimum. *The Mustard Seed Garden Manual of Painting* (1679) exhorts the student: "You must work hard. Bury the brush again and again in the ink and grind the inkstone to dust. Take ten days to paint a stream and five to paint a rock…. Study ten thousand volumes and walk ten thousand miles."

My arm ached already. In 1935 Chiang Yee gave a series of lectures for the International Exhibition of Chinese Art in London that became an influential book, *Chinese Calligraphy: An Introduction to Its Aesthetic and Technique*, introduced by the distinguished art critic Sir Herbert Read. Yee derides Western writing as mechanical and print as "visually lifeless," whereas Chinese calligraphy was like dance, "impulse, momentum, momentary poise, and the interplay of active forces combining to form a balanced whole." This stroke I was attempting should "rise slightly to the right end to give a touch of liveliness," he said. So much to think about for a single line. I wasn't dancing, I was plowing. I was lucky, though: if I were in China the teacher might make me balance a brick on the elbow of my brush arm to make sure my posture was correct. He would certainly chastise me by striking me with a closed fan.

"And don't hold breath," said Lisa. That was another thing. The character should emerge with the breath. I was holding mine as if I were trying to set a record for free diving. That and holding my brush in a death grip. Perhaps life was too short for calligraphy. "Oh no, you are doing well. You will get good," said Lisa. She was lying, trying to be Western. She did not do it well. The fraudulent Western encouragement has to be delivered with conviction, sincerity. We looked at my five pages of sausages and dumbbells. "Hmm," she said.

In the late 200s CE the aesthetic concepts of calligraphy were being refined in China, and the injunctions themselves soared into simile. Good brush technique "jumps like a stirring dragon, and falls as rain pouring from the sky"; a brush "may soar like a hawk in lofty flight, totally released, or [coiled] as delicate ornaments"; and when the writing is finished it is "as if a fair wind were skipping over cool water, creating endless ripples; examine closely, everything is natural." Unlike mine. Though my sodden paper did have endless ripples.

Calligraphy always had a moral as well as an aesthetic dimension. The famous eighth-century calligrapher Liu Kung-ch'üan declared, "A righteous heart conducts an upright brush." The work may be extolled for its qualities of line: "His dot is a falling rock, his line moves as a summer cloud, a hook is bent iron, a curve is a drawn bow," and also for what it reveals: "The sharp stiff strokes of his style indicate his obstinate character and determination to work out his own ideas." However, the critical effusions shift swiftly from images of the sublime to cruel invective: "In his style one can discern the loose flesh and easy manner of a fat person." Professional jealousy taints judgments—one critic said of Liu Kung-ch'üan (he of the righteous heart) that his writing looked like a "meat dumpling that is thickly covered with dough." He had my sympathy. I wondered what my sausage strokes revealed about my bad character. Chiang Yee writes that "many a friendship has been made or strengthened by admirably written letters." At the very least I would have had no friends.

II

I'd seen my first ink cake at the Metropolitan Museum in New York, in a display of the great ink makers of the Ming dynasty, Cheng Chün-fang and Fang Yü-lu. To me it looked like a fancy hockey puck, black with a design etched into the top. While Western ink remains a utilitarian product, Chinese ink makers long ago discovered that they could mould the ink paste into different shapes and embellish the surface with engravings. Little more than soot and glue, these cakes and inksticks became luxury commodities, markers of wealth and cultural status, works of art that could be used to gain favour with an emperor, and that he in turn would bestow on those he wished to honour.

Intrigued, I looked up a 1930 *Metropolitan Museum Studies* article by Wang Chi-chen, in which he noted that the best inksticks were made in the city of Shexian (SHUR-She-ann), deep in the hills of Anhui province. Wang also mentioned, in a tantalizing footnote, a famous quarrel between the two ink makers, precipitated by Fang "taking unto himself an ex-concubine of Cheng's." Clearly I had to go.

Ink cake, the size of a hockey puck, with coiled dragon.
Lao Hu Kaiwen ink factory. Photo courtesy of the author.

Yet research goes nowhere without the generosity of guides. One of our Chinese graduate students, Leilei Chen, tracked down the Lao Hu Kaiwen ink factory in Shexian. This factory made inksticks the traditional way, by hand, using designs that dated back to the Ming dynasty, and, after numerous phone calls from Leilei, they agreed to give me a tour. Even today the area is difficult to reach, a five-hour drive south of Nanjing on winding roads. This isolation is in part what made it great. The mountainous terrain kept farming minimal, so if you were going to make money it would be in trade, and there were good routes on the rivers going out—but that doesn't explain everything. Shexian is hardly the only hill-bound place in China. Across the mountains to the east was

Suzhou, a centre of the book trade, a city known for silk and style. Was there something about the beauty and insularity of Shexian that attracted a certain kind of merchant, one interested in beauty and culture as well as money?

I flew to Beijing, rocketed down to Shanghai on the high-speed train, and took the slow line west along the Yangtze to Nanjing. There I met Professor Yao Yuan, stylish in her Burberry overcoat and lyrical English (she translates Ondaatje and Yann Martel), improbably graceful in the frozen and rundown offices of the Foreign Studies division of Nanjing University. She had offered to help manage my trip to the factory. I asked her to write down the characters for her name in my notebook, and then I tried to reproduce them. "You're drawing!" she laughed as I dragged the strokes with my broad-nibbed fountain pen. I felt like a ham-fisted kid with a crayon. I told her I was worse with a brush. Still, the materials fascinated me.

Yao Yuan had arranged a translator. "Mr. Chi is a graduate student. He is very reliable," she said. In walked a young man in a tan jacket over a white shirt and dark-blue V-neck sweater, wire glasses, hair cut in a fuzzy flat-top. Very serious. Very nerdy. Perfect. Mr. Chi would arrange transportation. He investigated and found the bus to be twice as swift as the train—two and a half hours instead of six—and it left at seven p.m. instead of eleven p.m. Wonderful. We'd be there in the early evening, maybe have a late dinner and stroll about.

Then things started to unravel. He got into a shouting match with the cab driver on the way to the bus station, claiming he was taking a longer route. At the bus station when he went to validate the ticket, the woman laughed and then shrugged. Much talk followed. Had we missed it? Was it full? The talk went on. A bump on Mr. Chi's temple started to throb. The woman called another woman over. She smiled when she heard the news and looked at Mr. Chi as if he were a confused five-year-old. This was not going well. Finally he turned to me, anguished, about to weep. "They told me on the phone that the bus only took two and one half hours. Now they say it takes five and one half hours."

"So we'd get in about one in the morning. That's okay."

Sometime after one a.m. the bus lurched to a halt at a deserted intersection. The driver shouted something that sounded like Shexian. A single cab with a sinister-looking driver was waiting under the lone street light. Mr. Chi, who'd been on his cell phone constantly, explained that there had been a change of plans. The factory driver would not be meeting us. We would have to take the sinister cab. The responsibility of caring for the foreign professor was going to give Mr. Chi a heart attack.

The cab was fine in the end. The hotel was freezing, but all as expected, and in the morning the promised minivan took us to the factory. The buildings looked to have been built in the 1930s, though the original factory goes back much further. Pine trees lined the courtyard, and across the road a terraced hill rose. The driver ushered us up the steps, past an elegantly gnarled tree in a pot, and into the manager's office. The walls were white with black smudges and it was cold. The rickety fan on the wall and the tired-looking overhead fan indicated it would be hot in the summer. The manager offered us tea, and accepted my card and my letter of introduction presented by Mr. Chi. Soon the manager's son, Mr. Zhou, a young man in jeans and a sports jacket, appeared to take us on the tour.

We made our introductions and then he led us past a low flat cart loaded with black discs—raw ink, though I didn't know it yet—and into the first room.

Whack! Whack! Whack!

What hit me first was the pounding. A slight man in a blue turtleneck was kneading the balls of ink, rolling them into sausages, spinning them into balls, leaping onto his toes to put his whole body weight into it, dancing with the ink cake, flattening it, spinning it up again and flattening it, over and over.

Whack! Whack! Whack!

Rollers on one side, whackers on the other, and down the centre of the room a row of what looked like low round barbecues roasting old black overcoats. In fact they were covers to keep the big cakes (imagine a pancake two inches thick and all black) warm and malleable before it was their turn to come under the mallet.

Whack! Whack! Whack!!

These men really put their arms into it. Before a cake could be rolled it had to be pounded. They'd whip one out from under an overcoat and flip it onto a steel anvil and then grab a short-handled sledgehammer and … *Whack!* I don't know what I had expected— some peaceful kneading of soot like home bread making? Certainly nothing this violent. The serene inksticks of the study started with these explosive movements in the shop. As with printers, you would not want to scrap with these guys.

The simple clothes of the workers (no corporate uniforms) were contemporary but could have been from any era; no one in the room wore a mask, or safety glasses, and the only ventilation was the window. The men were perched on long flat poles, and the ink blobs were being pressed one by one in the moulds at the end of the stick. I asked how long each one was pressed for and Mr. Zhou said, "At least five minutes." This seemed very unscientific, but then so did the use of a padlock in the weigh scale on the first man's bench. Did they all have padlocks? They weighed each blob, made sure the scale went into balance, yet they all seemed to be using something different to measure the weight. Had they figured out a weight and then found articles to fit? The whole place seemed a mixture of precision and slapdash.

I say "seemed" because I didn't know what I was seeing. The man in front of us was trimming the edges of octagonal ink cakes, taking off the little tendrils that were left after he took the ink out of the mould. Mr. Chi had been talking animatedly with Mr. Zhou for five minutes. I asked what he was doing. "He is, um, repairing the ink," said Mr. Chi, and then he pointed to the scissors and knife. "These are tools," he said. I tried not to sigh. Mr. Chi was probably encountering terms unfamiliar to him in Chinese—he was, after all, a master's student writing a thesis on Chinese-American literature, arguing that it was not, as was commonly thought, based on a Chinese culture but on an American culture with Chinese elements. I'm sure he was one of the best students in the program at Nanjing, and it's not as if our graduate students would do well touring a foreign visitor around a print shop from Shakespeare's time. They wouldn't know a tympan from a platen, much less have it on the tip of their tongue in a foreign language.

At a table near the door, set off from the men hammering the black cakes, a man ground red pigment. He wore a sweater vest over a blue turtleneck. Where the others had sledgehammers on the stone stumps beside their tables, he had a wooden box with a bowl full of red powder set into it. He was working a long-handled pestle, and quietly rolling, not leaping and pounding. Vermilion red had long been used for stamping seals; a sacred colour and material, it had once been reserved for emperors. Herbs were blended in slowly with castor oil or cabbage-seed oil to keep the seal impressions bright for centuries. I wondered if the status of the man who ground the red was higher than the others, and if the mercuric sulphide was dangerous. It was too noisy, and the question too complicated to ask. I'd read that the best ink for printing in red was a mixture of vermilion and red lead, boiled in water with the mucilaginous root of a plant called *po chi*. I didn't know what *po chi* was, but I loved *mucilaginous* (which I'd encountered in Joseph Needham's vast *Science and Civilization in China*); the word derives from medieval French for "musty juice." Nevertheless, it all sounded toxic.

On the way out I saw a pan of beads and a pan of clear liquid. "This is glue," said Mr. Zhou. Carbon is reluctant to combine with other materials, so a binding agent is crucial. In China glues were traditionally made from a variety of animal remains, including raw hides, bones, and fish skins. The remains, whatever they were, were boiled up, strained through silk or cotton to remove the lumps, and then allowed to congeal until needed. I'd read that the best glue is made from the antlers of a deer. "Too expensive," said Mr. Zhou. "We use pig, cow, dog." Poor Rover.

Mr. Zhou took us up to the drying room, where the ink cakes were laid out on racks like buns in a bakery. I put three of the lozenge-shaped sticks on a rough wooden table in the sun to photograph them. Black, green, and yellow, with a long-tailed dragon running the length of the stick, they were impressive even while still unpainted. I counted the toes of the dragons—four, which meant they were *mang*. The five-toed horned dragon, the *long*, was reserved for the emperor, and in the old days if you had a gown made with five-toed dragons on it you risked being executed. Only the highest members of the imperial family were allowed to wear those designs. These dragons, with legs out and claws splayed like a cat on slippery tile, were called dancing dragons. The round ink cakes had coiled dragons: *p'an lung*.

Upstairs you move into a world of quiet and colour, the loud whacks replaced by the soft *shhh shhh shhh* of young women polishing the ink cakes brought up from below. They were leaning over the cakes, chatting, dressed in jeans. By the windows, more mature women in traditional clothing painted the engraved scenes. The colours were radiant, the women deft. I joked that it was ever thus—the men do the crude heavy labour, the women do the skilled delicate work. Neither Mr. Zhou nor Mr. Chi commented. Had I made a faux pas, or simply stated the obvious?

After the polishing room, Mr. Zhou took us across to another upstairs room. This was the most placid one of all. No smell, no noise. Only two people, each working by their own window. Easy to forget in the excitement of the whacking, the grinding, the painting, the polishing, these were the people who made the designs: the mould carvers. The blue curtain was knotted to let in more light—it came in over their left—and on a string between the windows sagged a towel and a half-open umbrella.

One man, one woman, equally cold. Both wore jackets and worked with the windows open—for the extra light; there was no heat, so there was nothing to be gained by keeping it closed. On the floor between them lay a heap of used wooden moulds. On their tables sat the fresh bright wood for carving. Fine shavings littered the table in front of them, among the toothbrush and the several thin carving tools. On the other walls hung a few worn reproductions of scroll paintings—the famous landscapes of Huangshan (Yellow Mountain), an interior scene with a bamboo plant. I asked if these designs were there to inspire them, or to be reproduced. "Yes," said Mr. Zhou. We moved over to the unoccupied table, where he picked up a picture of a peacock and then its raw carved mould beside it for us to see, a marvel of intricate feathers in wood.

Mr. Zhou took us through to a big room labelled *Shopping Centre*, where the work was showcased and customers could come to make purchases. The space was all dark, except at the front where light streamed in, glancing off the yellow walls and soaking the red tablecloth in a deep hue. Arranged on the table were the Four

Treasures of the Study: paper, brush, inkstone, and ink. Each of the highest quality. There were other small objects on the desk—water droppers, brush rests, ornamental rocks—all chosen for aesthetic appeal as well as utility, so that the calligrapher's eye might rest on them for inspiration before the next stroke. I wondered if a grimy keyboard, a smudged screen, and a desk cluttered with yellow stickies and takeout coffee cups affected one's prose.

The desire for scholars to surround themselves with beautiful things originates in the belief that the objects are receptacles of "vital essence" (*qi*), which strengthens your vitality and lengthens your life. According to R.H. van Gulik, undeveloped deer horns are supposed to be rich in *qi* (so perhaps this is why it was regarded as producing the best glue), and the pine tree is not only a symbol of longevity but also rich in *qi*: Taoist adepts are said to have subsisted for years on a diet of pine seed. When writing, the scholar will from time to time shape the tip of the brush by licking it—and thus partake in the magic properties of the ink. *Ch'ih mo*, to feed on ink, became an expression for painting and writing.

There were sumptuous brocade boxes in yellow and red, blue and gold, with little gold clasps, and inside, nestled in satin, inksticks of dazzling design. Bright golden dragons, red and gold cocks. Sticks as long as a man's forearm, so richly coloured, so ornately wrought that you forgot they were ink. These must be the ones fit for an emperor, ones no one would ever use. "Oh no, they use," said Mr. Zhou. My favourite was a box with eighteen scenes, nine cakes on each side, of Huangshan, source of the best pine for ink. Blue and gold dominated, blue or green for trees, white for the little cottage nestled under them at the foot of the mountain and for accents on the mountain spires. The next day, riding up Huangshan in the cable car, I would see the exposed patches of granite, shiny, almost white.

To use a big inkstick you need a big inkstone. There were a few the size of a hand, but most were at least as big as a dinner plate. Some were massive blocks of slate, and sold for $10,000. Like Inuit sculpture, the best ones drew you in.

One February I spent two weeks in Yellowknife writing while Hsing worked at the local hospital. She was on call, so I was alone most of the time in a log house on Back Bay, my only companions two sculptures. "The guy with the big eyes and gapped teeth is by Uriash," Hsing said. "The dancing bear is by Judas, from Gjoa Haven, one of the last generation of carvers who grew up on the land." They spooked me a little—I was used to my mother's smooth ptarmigans—but by the second week they'd become welcome companions. I scoff at New Age crystals, but those northern sculptures had some kind of power, and I felt a similar force with these inkstones. All were skilful, many intricate, but several were daunting, mysterious, and the high, cold, quiet room resonated with their presence. My favourite was a lozenge-shaped stone of greenish-brown hue with two water buffalo lying down, curved contentedly, as if cooling themselves in a muddy bog. When we came back later and customers were sifting through items, passing calculators back and forth, bargaining, the magic was muted, as if the stones had drawn in, waiting until the crowd had passed.

III

As we crossed the courtyard, I looked into a black doorway. A light bulb glowed feebly through the murk. A man with the face of a pirate was looking out. I smiled and started to take his picture. He smiled and moved slowly back into the smoke billowing out of the doorway, gone like the Cheshire Cat's fading grin. My photograph shows only an empty doorway. I caught up to Mr. Zhou and asked about the room. "Not allowed," he told me. I had seen the origin without realizing it. Smoke. I'd read about the process, but now, instead of imagining fine soot carefully sifted by workers with white feathers, I thought of the tar globules clogging the air sacs of the pirate's lungs.

Ink, as Steven Smith had told me in Texas, is a colloid—a mixture in which the particles don't dissolve but remain particles—and the blackness of Chinese ink is affected by the way the glue "delivers" the carbon particles to the page. The grains coagulate into blob-like structures, forming layers on top of the silk or paper, while some of the finer grains penetrate to the back of the paper. So our perception of the ink will be determined by many things, including the quality of the soot, the grinding method, the type of brush, and even the way the light is reflected. The great Chinese ink painters felt no need for colour, for they could command a range of grey tones as broad as those in an Ansel Adams photograph, where you feel rather than see the colour.

But colour is tricky. There are two basic kinds of carbon particles for ink: soot from pine smoke, and soot from burning vegetable oil, or "lampblack." And there are two basic blacks: those tinged violet/blue and those tinged brown/red. The ink maker could enhance these qualities with additives—powdered ox tongue or the bark of the pepper plant—but what really determines the colour of the ink is the size and uniformity of the grains, which determine their light-absorption characteristics. Blackness increases with the smallness of the grains of soot. Oil soot is supposed to be blacker than pine soot because it produces smaller grains; it also produces an ink with a reddish hue, since the scatter strength of light increases with short wavelengths. So the properties of the ink may be determined more by the granular structure of the original smoke than by the hue of the particles or the additives. Ink as solidified smoke—the metaphoric possibilities were endless.

During the Ming dynasty they made their inksticks from pine smoke. I liked the idea of pine, of ink deriving from some craggy tree on a mountain cliff. Much more romantic than boiling down tung oil. Before cutting the tree down you must eliminate the resin, because even a small amount left in the wood will reduce the free-flowing quality of the ink. To get rid of the resin, says Ming author Sung Ying-hsing (1600–1660), "a small hole is cut near the root of the tree, into which a lamp is placed and allowed to burn slowly. The resin in the entire tree will gather at the warm spot and flow out." Only then do you cut down the tree and saw it into pieces. The draining procedure reminded me of making maple syrup, and the burning facility of an Iroquois longhouse. Sung describes it as a rounded chamber of bamboo "resembling in appearance the curved rain-shield on small boats, [with] ... a total length of more than 100 feet."

The pine burns for several days, allowing the smoke to carry the carbon particles through the chambers. Then the chamber is allowed to cool, and workers go in to scrape out the soot, though *scrape* is too coarse a word. The eighteenth-century historian Jean-Baptiste Du Halde (who compiled reports of Jesuit missionaries and never travelled to China himself) recorded how the workers "with a goose feather gently brush the bottom, letting the soot fall upon a dry sheet of strong paper. It is this that makes their fine and shining ink." The soot from the last one or two sections, wafted the farthest from the fire, is the lightest and is used for the best ink. Soot from the middle section is of middle quality and goes into ordinary ink. The coarse soot from the first one or two sections may be ground by printers and used for books, or sold to lacquer workers and plasterers who will use it as black paint.

Good ink not only has lustre; it has spreading power and adds smoothness to the brush movement. I thought of the cheap paint I'd bought in student days to brighten the nineteenth-century walls of my apartment in Kingston, Ontario. Why would you pay fifty dollars a gallon when you could get the same amount of paint for ten bucks? I soon found out. Bad paint dribbles or clots, frustrates your every stroke, and in the end doesn't cover the wall. So too with calligraphy. Zhao Mengfu (a calligrapher so famous he's had a crater on Mercury named after him) says, "It is as impossible / for a good calligrapher / to write with bad ink / As for a fast horse / to win in the mud." Poets wrote odes to fine ink. Here is Xue Tao, a Tang dynasty poetess (c. 770–832) for whom ink is mysterious and erotic:

> Old pine burned forming light charcoal flowers.
> The exquisite ink-making skills of brother Li.
> How describe the deep, cool shining colour?
> Darker than the fair lady's hair, a crow flying in winter.

Li refers to Li Chao; his family name was Xi, and *Li* was bestowed as an honour by the emperor Li Yu, who greatly admired his ink and made him the official ink maker.

Hsü Wei wrote the charming "Two Fish by a Willow Embankment":

> A row of willow trees, almost green for the spring,
> and two fish, lively, ready to jump.
> Maybe they will jump, and land in my inkstone—
> it happens all the time;
> I'm only afraid that they might splash ink on my clothes.

He was known for his "utterly individualistic, expressive style of quirky brushwork," a style that influenced later generations. Alas, he was more than quirky in his personal life: a talented poet, playwright, and painter, he tried and failed the provincial civil service exams eight times, and died impoverished. He murdered his third wife, was sentenced to death but released from prison after seven years, and became known for his gruesome suicide attempts, which included drilling his ears and smashing his testicles.

For Chang Yü, in his "In a Book-Box I Found the Lost Manuscript of a Poem Sent to Me by the Late Kao [Ch'i]," the perfume of the ink becomes a reminder of the dead beloved:

> Burning orchid-lamps, we invited the moon to join us;
>   drinking wine, we plucked the strings of our lute.
> Who would have thought that for another evening of joy
>   we would have to wait for thousands of years!
> Now your wandering spirit is far away in darkness,
>   and only cold words are left, in your own hand.
> The dusty ink still gives off a light fragrance,
>   the paper is torn, but still has lustre.

Ink and love run together. Zhao Mengfu, who likened writing with bad ink to riding a horse in the mud, was famous for his loving marriage to poet-painter-calligrapher Guan Daosheng (1262–1319), and for their exchange on the subject of concubines. Zhao wrote his wife a poem in which he said, testing the waters,

> I am a xueshi, and you are my wife.
> Haven't you heard that Tao xueshi has peach leaves
>   and peach roots

and Su xueshi dawn clouds and evening clouds?
If I marry a few concubines, they won't threaten your status.
You're almost fifty, so simply occupy Yu Chun Tang.

A *xueshi* was a high-ranking scholar who worked closely with the emperor; the *peaches* and *clouds* are metaphors of concubines; and *Yu Chun Tang* is the position of a wife or matriarch, but I think the drift is obvious even without commentary. We see here the eternal obtuse hopefulness of men; this is an argument that, even a thousand years later, you could imagine a certain type of man trying to make:

1.   Having a concubine is appropriate to my rank.
2.   Both Fred and Bob have one.
3.   A few peaches on the side won't change your position in the household.
4.   You're old, so be satisfied with being a matriarch.

The argument didn't work even in the thirteenth century. In reply, Guan Daosheng wrote her poem "Wo Nong Ci":

You and I
have too much passion.
Where the passion is, is hot like fire.
I knead a piece of clay into a you
and a me
then smash them
and mix them with water.
Again I knead it into a you
then a me.

There is you in my clay,
and me in your clay.
I'll share your quilt while we live
and your coffin after death.

Thus she expressed her love, and her implicit complaint about concubines, though they are not mentioned. Zhao, no fool, never spoke of them again. Guan's is a much better poem, still powerful, even in translation. It's a beautiful expression of passionate love, and a young friend of mine read it at her wedding (I wondered if she knew it was a cautionary tale for her man). And, having watched inksticks being made, it was impossible not to see the process of calligraphy implicit in the work: Guan was breaking the fired soil of an inkstick, mixing it with water to make with her brush "another you and me" in the poem.

For Li Po (also known as Li Bai, 701–762, as famous in China as Shakespeare is in Western culture), ink is a precious, intimate gift:

Soot made of Shang-tang Mountain green pine,
Mixed with cinnabar powder of I-ling,
And orchid oil and musk, a precious ink is made,
Its glaze shines so luxuriously that one is afraid to use.

The servant boy with two coiffures brought it in,
Wrapped in a brocade sack, carefully on his arm;
With this gift from you I am going at once to the Orchid Pavilion.
When inspiration comes upon me, I shall write happily with
my brush and your ink.

Cinnabar is mercury sulphide, the bright red powder, and the Orchid Pavilion was the site of the most famous literary gathering in China, three hundred years before.

These six poems celebrate poetic art and the materials of the craft, from the pine soot to the inkstone, to the scent that evokes the lost companion, to calligraphers and concubines, to ink itself as a metaphor for lovers, and finally, with Li Po, to a drawing together of pine, musk, love, and inspired calligraphy. It's easy to assemble such a sequence of Chinese poems, but it would be impossible in the Western literary tradition. In his sonnet "Shall I Compare Thee to a Summer's Day?" Shakespeare puns on how his "eternal lines" will give life to his beloved, but writing materials are never the subject of his poetry. Somewhere in the huge body of Western verse, lyrics about quills or pen knives must exist, but these would be mere curiosities, and I know of none devoted to the components of ink. There is no "Ode Upon a Gallnut"—even though, as we shall see, the humble gallnut creates an ink to outlast time. But in the Chinese tradition, the craft of making inksticks, the decorative art associated with them, and the literary arts were all intertwined.

The Chinese tradition also celebrates the confluence of ink and wine. The Orchid Pavilion event was a drunken revel. In 353 CE, forty-two literati gathered at the Spring Purification Festival in Shaoxing. The preface to the poems tells us that the sky was bright, the air refreshing, the gentle breezes soothing. A stream ran through the garden, on either side of the pavilion, and the scholars ranged themselves along the banks. Servants floated cups of wine down, and whenever one stopped, the man closest to it had to drink it down and write a poem. Of the forty-two poets present, only eleven managed to compose the two poems required of them. Fifteen managed to compose one poem, while sixteen poets did not compose a single poem, no doubt, says the historian Shih-ch'ang, out of regard for their future reputation (so maybe drunken inspiration doesn't work for everyone). Yet how could they know their picnic would be famous seventeen centuries later? Li Po could invoke the gathering four hundred years after the fact as a touchstone for poetic inspiration and drinking. Dubbed one of the "Eight Immortals of the Wine Cup" by his friend Du Fu, Li Po is famous for his poems celebrating the joys of drinking (even hangovers—see his "Waking from Drunkenness on a Spring Day"), and with his death he became immortalized in legend: he died, the story has it, while reaching from a boat to embrace the reflection of the moon.

I was about to encounter that tradition of ink and drink.

IV

As we walked back to the office Mr. Chi said, "The manager wants to have lunch with you."

"Sure," I said. It was only eleven, but I was getting hungry. We sat in the office and were given paper cups of tea again. Mr. Zhou asked if I had any questions. I consulted my book and was reminded to ask about pine soot versus oil. He said they used cedar and vegetable oil, and that the vegetable oil was better—what was used in the smaller, less ornate sticks he'd given me. The proportion of glue to soot? "Sixty percent soot, forty percent glue." Was that true for all the sticks, I asked, or did it vary for the more expensive ones? "Oh yes," there is more soot in the more expensive ones. How much? Mr. Zhou couldn't say, just that they added precious minerals and reduced the glue in the better inks.

What about the sound as a test? This was hard to convey. I made the motion of rapping on the table with my inkstick. "Tone.

Sound," I said to Mr. Chi. "How does the sound tell the quality?" He said something to Mr. Zhou, who nodded and went out.

"The manager wants to have lunch with you and it will be soon," said Mr. Chi.

"Yes. I think it will be interesting," I said. Mr. Chi said nothing.

Mr. Zhou brought back two of the hexagonal inksticks and clicked them together. They made a high clear sound, like castanets. "Good," he said. Then he held up two cigar-length sticks and hit them together. They sounded like two hockey pucks bumping into one another. A dull rubbery thud. "No good."

He took the inksticks back out to the room next door, and Mr. Chi said, "Soon we will go to lunch!" as if this was our last chance to not do lunch.

"Yeah, yeah. Fine. That's great." Why would we turn down a free lunch? Maybe I could learn more about the factory, how the designs were handed down, how it had stayed in the family. I was really hungry now, and getting a bit testy. Mr. Zhou came back. The manager was not around.

"Now we go," said Mr. Chi mournfully.

The driver took us back to the centre of town, to a restaurant on a corner across from the big Ming gate that is the centrepiece of Shexian. Girls in red and gold *qi pao* greeted us. One of the girls led us upstairs—she was wearing the thick flesh-coloured hose that seems to be required in China; more like long underwear than stockings, it completely negates the eroticism of the form-fitting *qi pao* with its long slit up the side. Maybe that's the idea, or maybe it's just that it was cold; one of the girls was wearing jeans underneath her dress.

We entered one of the smaller private dining rooms. There was our host at a round table for thirteen, already with four of his friends, all smoking furiously. The quiet, almost severe man I'd met that morning had been transformed into a loud, gregarious drunk. He waved us over, and when I started to sit a few chairs away from him he roared something, grabbed me—not quite by the scruff of my neck but by the folds of my jacket—and pulled me over to his left side. He made one of his cronies vacate the seat beside me so that Mr. Chi could sit there to translate. Alas, I needed no translation for this. I knew we were in for the Business Lunch. This was what Mr. Chi, with his thrice-voiced announcement, like the warning that comes three times in fables and is always ignored with dire consequences, had hoped to avoid.

In his memoir of teaching in China, *River Town: Two Years on the Yangtze*, Peter Hessler talks about how these lunches become aggressive acts, with one weak link singled out, forced to drink until he falls down and vomits or otherwise disgraces himself. At the same time I knew that you weren't supposed to take a drink without toasting someone. I opted for wine instead of beer—a small glass instead of the 600 ml bottles they were bringing out—and with the first few toasts clinked glasses and took a sip. This was going to be fine.

Then my host roared and one of the brocade girls brought him a bottle of clear fluid with a spigot. He offered me some; I indicated no thanks, pointing to the wine, and he poured some for his friends in little thimbleful glasses. They stood, he shouted a toast, and they tossed the glasses back. We went on eating, the centre Lazy Susan full of duck, chicken, beef dishes, shrimp on a plate, spinach, bok choy, snow peas.

The host reached over and poured me a thimbleful. Alarmed, Mr. Chi tried to intervene. "It's okay," I said. I'd do this one. My host looked pleased. We said cheers and something in Chinese and knocked it back. Kind of like fruity-flavoured vodka with an oily aftertaste. Not bad. Not good. Every culture has a variation (vodka, tequila, ouzo), a powerful clear liquid that's basically alcohol and whose sole function is to get you hammered. Still, I'd done my duty, and we went on eating.

My host started to pour another. The quick Mr. Chi put his hand over my glass and said something earnest to our host. The host said something harsh back. "It's okay," I said. "I can do this." We stood up, there was some speechifying, and I drank it off. Not bad, but not something to mix with wine. I put more rice in my bowl and applied myself. I wanted to have something bland and absorbent in my gut, not oily, spicy sauces, if this was the way things were going.

When my host picked up the bottle again, Mr. Chi intervened and I said, "Tell him no thanks, that I have the wine," and pointed to my glass. Apparently this was okay. A brief speech, my host knocked his back; I took a sip of my wine and put the glass down. Now this was not okay. My host leaned in close to my face and said something, pointing to my glass. I took another sip. He pointed again. The entire table was looking at us now. He made it clear that for the toast to count I had to chug the glass of wine. His glistening face was still close to mine. Everyone was smiling. What the hell. The wine wasn't too bad. I glugged it back in four swallows.

"Okay, that's enough," I said to Mr. Chi, but almost immediately my thimble glass was full of whisky again and my host was rasping in my face. I could feel the wine and whisky mingling in my brain as well as my stomach now. Getting sozzled. Getting to that point where you think, Well, why not tie one on? But doing shots of oily whisky at eleven-thirty in the morning is not my idea of a good time. I also knew that the mixture would produce a devastating hangover, and I didn't want to spend the afternoon dry heaving and the next day with a splitting headache instead of going to Huangshan, the most beautiful mountain in China. Besides which, I did not like this guy. Yes, I was honoured to be honoured, but Mr. Chi had been explaining that this was not the custom in my country. My host didn't care. At some point I presented him with a bag of maple candies, which I think he made a joke about to the table. Mr. Chi gave me only very brief, sanitized translations of long exchanges, translations that did not explain the laughter. I also noticed that the others around the table were allowed to take small sips of their drinks after the toasts, to go on drinking in a normal manner. Maybe I should just get drunk and chalk it up as my participation in Chinese culture. I knocked it back. I have always been a cheap drunk. Two drinks and I'm lightheaded. Three and I'm leering at the waitresses. Four and the world sickly spins.

Now my host was snarling at me, and at Mr. Chi who was leaning across the table, one hand covering my glass and the other gesticulating as he insisted whatever he was insisting, that Canadians didn't do this, that we'd had enough, and so on. Mr. Chi was a fierce bargainer; I'd already seen him in action, and this was quite a contest. But then my host shouted something and the whole table rose. The one woman in the group looked at me with amused sympathy, but she too raised her glass. They all had their glasses raised. Okay. I stood up too. My host roared something and we all drank at once. I held my glass up, as we all did, and then picked up my glass of tea, drank, and let the whisky in my mouth flow into the tea.

I sat down. The other shots of whisky and wine had found their way through the pad of rice in my stomach and into my bloodstream. I was glad I'd fudged that last one. People were starting to leave. This was, after all, just lunch. If it had been dinner there would have been no break, no chance to escape, and I'd be carted off to karaoke. My host got up to see someone to the door. "Let's go," said Mr. Chi. As we reached the door on one side my host was already back in his chair, the table empty. He roared and gestured for us to sit down again. Great. One-on-ones with the bottle of whisky. Mr. Chi said no, we had to go, but thank you very much. I said thank you.

My host sat there and we headed down the stairs. We reached the street, and he appeared moments later. "The driver will take us back," said Mr. Chi. When I turned to thank the manager again he was deep in conversation with one of his cronies, and so with no formal goodbye we were bundled into the van and taken back to the hotel. At the hotel I found I wasn't that drunk, didn't feel ill, and thought, I should have had a few more shots. But a few more wouldn't have ended the game. Two days later in Nanjing, Yao Yuan said, smiling, "I could have told you that would happen. Your mistake was to start drinking at all. Because then when you say no they just think you're being modest. It is worse in the North. There you would not have had the option of stopping."

Mr. Chi had been very apologetic. "He just does not understand. He does not care about cultural difference."

"It's okay," I said. "It's more than just a Canadian–Chinese difference." Maybe I should have been grateful, should have gotten drunker and tried my hand at some poetry, embraced the tradition of drunken calligraphy, the legacy of the Orchid Pavilion. But I doubt it would have worked. It was not my tradition. Two pints and I can't even keyboard, and there's no auto-correct with a brush.

V

The next morning we hired a car to take us to Huangshan, the "Yellow Mountain," so-called because legend has it that the Yellow Emperor ascended to heaven from here in 2598 BCE.

This is the region that has inspired all those scroll paintings you see in Chinese restaurants in North America—fluted slabs with knubbly pine trees on top, maybe a waterfall and a little scholar's pavilion tucked in at the base. What was once a place of pilgrimage is now a tourist trap where the parking is extortionate and the fences sag with padlocks placed by lovers to signify their eternal, locked-in love, but all grumbling stops as the gondola swings you up through the spires. With my Canadian Rockies snobbery I expected pretty hills, but these are real mountains—they rise seventeen hundred metres from the valley floor, with granite pitches that would give any rock climber itchy fingers. The winds are harsh and the pines dig in with dragon-claw roots. Huangshan is like Machu Picchu, with lung-sucking climbs up endless fifteen-hundred-year-old stone steps, and then downhills swift and effortless—until your knees start to give. The pain adds another dimension to those restaurant paintings.

I tried unsuccessfully to take pictures without people and to imagine the mountain without the big television tower on top, yet the area was still sublime, beautiful, and fearsome. Mr. Chi had never been there, and he too was dazzled by the beauty and scale of the place. We rambled for a couple of hours and then stopped at the White Cloud Hotel. Mr. Chi taught me how to say "fried rice" (*chao fan*) and "dumplings" (*jiaoza*) so that I could order if I was in a restaurant with no picture menu. (I didn't know that a week later I'd be travelling alone out to the edge of Tibet, glad of this tip.) We talked about what it must have been like to climb these steep paths in early dynasties, gathering gnarly pine roots in a basket. Huangshan was so famous for its pine trees that one tenth-century ink maker moved to Shexian because of them. The compiler of *The Mustard Seed Garden Manual* says that pine trees "resemble young dragons coiled in deep gorges; they have an attractive, graceful air yet one trembles to approach them for fear of the hidden power ready to spring forth." Another writer likens these "young dragons" in gorges to a spiritual power emanating from the unconscious, a notion that seemed less fanciful now that we were among them.

Back in Shexian, as we walked across the bridge, the Christmas-style lights edging the buildings reflected in the water, Mr. Chi asked me what authors I liked. I told him I was reading Du Fu. "Du Fu?! He is one of China's greatest poets. The first poem my father made me memorize was by Du Fu. I did not want to. I had to recite it for him, and for many years I did not understand it, but now I do."

He went on, "I wrote a poem once, in English." I said that I'd like to read it sometime. Mr. Chi touched his jacket and said he had it with him. Of course he did. I said we'd read it in the restaurant. I hate reading with the eyes of the author upon me, especially those of an expectant poet, who over noodles explained how there was a shift after the first quatrain and a big change after the octet because the relationship with the girl had changed. T.S. Eliot was his favourite poet, he said, and so the references to the "hyacinth" and memory were to *The Waste Land*. He said that people now had no integrity. "In the old days people had more honour." (I would later meet the vivacious Holly, his Hyacinth Girl, who had played Isabella in an English production of *Measure for Measure*, fallen for Angelo, and shunted the earnest Mr. Chi into the friend zone.) I reminded him that Proust says all paradises exist only in the past, and told him the story of the love, betrayal, and murder that had intrigued me in New York.

Thousand-year-old tree, from Cheng's catalogue. Note the distinctive Chinese clouds (at first glance I thought the tree was underwater), which recur in the next illustration. From Wang Chi-chen, "Notes on Chinese Ink," Metropolitan Museum Studies 3.1 (Dec. 1930): 121; original in American Museum of Natural History.

Cheng and Fang,
Rivals in Ink and Love

I

Cheng Chün-fang (1541–c. 1610) was born into a rich merchant family, and at university adopted the lifestyle of an aesthete, composing poems and articles, neglecting his studies. He became obsessed with collecting inksticks. A classmate later recalled that Cheng spent huge sums on ink, and that he'd chided him about it: "You have a weakness for inksticks. Inksticks are going to grind you"—that is, instead of you grinding them, they will torture you—and in truth they did. They would ruin but ultimately redeem him.

Cheng knew he would fail the harsh civil service exams, so he bought a post in the Court of State Ceremonies; he'd be in charge of receptions for foreign dignitaries. However, he proved "haughty and quarrelsome" and was dismissed. Back in Shexian, he prospered in the ink business and took on Fang Yü-lu (c. 1541–1608) as an apprentice, teaching him the trade and supporting him with money and servants to develop his own business.

Fang too was the son of a local merchant, though according to Cheng he was also the son of a prostitute. In personality Cheng and Fang Yü-lu seem to have been opposites, the rigid moralist versus the passionate profligate. Fang's supporters agree that he spent money recklessly and loved pleasure, but insist that he was very generous to his friends. When poor, Fang devoted himself to study, and achieved such skill as a poet that he was invited to join the local literary society. But according to Cheng he had his father's libido; he spent all his money in brothels and wound up penniless. Worse, he had his eye on Cheng's concubine. A century later, in the 1600s, the story was still famous. Chiang Shao-shu writes, "Cheng had a concubine, who was very beautiful, of whom his wife was very jealous. Cheng went to Beijing on business, came home to find his wife had thrown out the concubine, and that Fang, who had always lusted after this concubine, had employed someone to get her for him to marry." Fang married her, but Cheng then sued and Cheng and Fang became rivals. Fang lost the case and the concubine. But Fang succeeded in stealing Cheng's ink designs: he published them under his own name in eight volumes, and though the outraged Cheng later published his own catalogue in twelve volumes, it was too late. Fang had cornered the market.

In the meantime, Cheng tried and failed the exams in Beijing again. On his return to Shexian he flogged a servant so brutally that the man died after nine days. Cheng protested that he'd been framed, that Fang had bribed a doctor to prescribe a decoction of rhinoceros horn that included arsenic. Cheng was never convicted, but he spent seven years in jail where he wrote furiously, pouring out a series of poems and articles that he would later include in his catalogue. Accusations circulated on either side—that Fang had tried to poison Cheng, that Cheng fed children to the snakes in his garden that he raised for poison to use in his ink—but all agreed that Cheng and Fang made marvellous inksticks. According to one commentator, "The literati treasure them like precious jades," and the best ones were inscribed *pu-ko mo*: Do Not Grind.

Some of the literati disapproved. One Ma San-heng complains, "Just the brightness of the gaudy outward appearance [of the inksticks] is like putting on brocade. It is said that a person should not use his looks to win his fame.... How has [the decoration of inksticks] come to be important?" The effect is pernicious, he says: these inksticks won't help anyone with their literary work, but now other ink makers dare not do without the elaborate embellishments. The function of inksticks, Ma laments, is no longer as "useful tools but as art objects." Whether the literati liked it or not (and most of them did), Fang and Cheng had altered the aesthetics of ink making.

And yet inner quality still counted. Cheng's supporters claimed that Fang "put smelly black dirt inside of his inksticks and sold them ten times more than the average." But "some of the inksticks were worn out by wind and then broken, and what was inside was exposed." The gentry began to doubt Fang's integrity. However, a commentator on Fang's side declared that the poorer Fang became, the more he concentrated on the calibre of his inksticks: "[He] was not concerned with the images, but concerned with truth (the substance of the inksticks)." So the inkstick becomes an emblem of character: a meretricious exterior may hide a corrupt interior. Whatever the truth of it all, I said to Mr. Chi, honour and integrity were not the strong suits of these brilliant ink makers.

II

In his travel book *News from Tartary*, Peter Fleming (brother of James Bond author Ian Fleming) writes, "In China your calligraphy is an important clue to your social status; it serves something of the purpose of your Old School Tie in England, but is of course much better fitted to serve it." That was in his 1936 account of travelling from Beijing to the Kashmir in India. I wondered if calligraphy still counted. Mr. Chi told me his father was a calligrapher and that he himself had studied calligraphy up until secondary school. "But after that I stopped. I had no time. No one does it after that because—you have heard of the examinations for the universities?—yes, we have to study very hard."

"So will calligraphy die out?" I asked.

"Mmmm. I think the young people still study it. Age ten. So it will continue." But at a ten-year-old level? Who would teach it? Calligraphy had once been required for court positions. I told him I'd read that the Chinese had great reverence for any paper with writing on it, that every district of a Chinese city has a pagoda—called *Hsi-Tzu-T'a*, or Pagoda of Compassionating the Characters—for the burning of inscribed waste paper, and that one would see old men with bamboo baskets on their backs picking up this kind of paper. I never did, and Mr. Chi said no, he'd never seen that either, though he'd heard about it.

In the 1500s, the Chinese concept of a scholar differed radically from the European. Montaigne (who was working on his first volume of *Essais* about the time Fang was trying to marry Cheng's concubine) didn't have to pass state examinations, and ability in art wouldn't have been expected. Chinese students began with twenty-five simple characters, working on big sheets with red-lined squares made up of nine smaller squares so that they could learn the characters' proportions and placement as well as the strokes. I remembered workbooks with dotted lines between the solid lines to help us achieve the right proportion of bowl to ascender on lowercase letters like *b* and *d*. But we had only twenty-six letters, simpler than all but the very simplest Chinese characters, and at age nine, when we were being allowed to use straight pens, Chinese children would have been expected to have mastered the *Primer of One Thousand Characters*—250 lines in which the same character was never repeated. By age fifteen they would have worked through the *Analects* of Confucius as well as other classical texts. (My students from China tell me they still do.)

Nigel Cameron in *The Chinese Scholar's Desk* captures this gruelling process. For the District Examination a cannon was fired at four a.m. to wake the examinees from their fretful sleep. A second cannon set them stumbling by torchlight through the pitch-black streets, unfamiliar to those away from home, carrying the Four Treasures of the Study, along with water and a lunch, to the big examination hall in the district's main town. Candidates took their seats, the officiating magistrate entered in ceremonial robes and personally locked the doors, and the first question was pasted on a placard in large characters. Questions were sometimes so obscure that all the candidates were baffled. When that happened the examiners called out loudly, "We have outwitted them!" and asked a different question.

The answer had to be in the form of a poem written on a set theme, rhyming and in five-character lines. (When Mr. Chi told me he'd produced a sonnet with echoes of T.S. Eliot, I recognized the tradition.) When it became too dark to write the candidates departed, and for the next three or four days the magistrate and his secretaries would be locked in the hall until they had graded the last paper. The names were posted on the wall of the magistrate's headquarters, meaning that the triumph or disgrace you'd brought to yourself and your family was visible to all. A red dot appeared against your name if you were successful; no dot if you'd failed.

One Ming poet complains, in terms familiar to our century, how "My eldest son travels the world, / Barely managing to make a living. / My second son claims to be studying, / Yet only desires the pleasures of wine...." (It seemed to me that in China children were everything—yet always a disappointment. Is it significant, I asked Hsing, that the word for *mother* and the word for *scold* have the same pronunciation, just different tones? In family life it's as if your good deeds are written in water, your failures etched in stone.) That Ming-era poem was inspired by one written a thousand years before, Tao Qian's lyric on his five sons' shortcomings: "not one of them fond of brush and paper." An affinity for brush and paper, for calligraphy, was the mark of responsibility, handling of ink a measure of moral worth.

For even if you passed the district exam you could not rest. Successful candidates advanced to the provincial capital for the Provincial Examination. Here, instead of an exam hall, you filed into a large walled enclave with rows of huts made of bricks with three walls and a roof (often leaky); there was no fourth wall, they were open to the wind and rain. You brought bedding, a portable stove and food, your inkstone, ink, brushes, and paper, and a chamber pot; this would be your house for three days and two nights. The gates were locked, and soldiers were stationed to suppress any disorder.

The tension was terrible. Imagine if the LSAT or MCAT exams were all day long for three days and you had to write them in a beach cabana in a stadium and you had to bring your own Coleman stove and your own pot to pee in—and then when you found out you'd passed you had to do two more such sessions before you could get into law or medical school. Of course you'd be slightly crazed. Candidates often suffered hallucinations, including that of young girls gliding between the huts.

But you still weren't done, because every three years the Palace Exams took place in Beijing for the highest level, the *jinshi* degree. Candidates assembled in Tiananmen Square, a daunting space where you feel exposed and vulnerable, facing massive walls and the impassive Southern Gate. Then the doors swing open and you enter the heart of the kingdom. These were the exams Cheng had failed. Sometimes, if the emperor wanted to keep certain people from advancing, he'd fail the whole class—as happened to the esteemed poet Du Fu. I'd always envied those idyllic pavilions in landscape paintings, but they were cages of stress and anxiety. Scholars read in their pavilions until their hair turned white, their eyes misted over with cataracts, their stomachs corroded with ulcers.

The system that had kept calligraphy alive, forcing millions to learn it to a high level so that as a byproduct a few stellar exemplars emerged, was now causing it to decline. Would calligraphy become an elite pursuit, like playing the lute or the harpsichord? I wished Mr. Chi senior lived in Nanjing so that we could have a conversation about these things. Chinese calligraphy is far removed from Western notions of writing. It's not a hobby, or an arcane pursuit. As Nigel Cameron says, no one in the West ranks penmanship with great painting, but the Chinese have always done so. Yet the analogy isn't exact: good painting would never get you a job in the civil service. Calligraphy and professional competence are allied. A professor at a Shanghai university told me that calligraphy could still be a factor for hiring committees. The most dramatic marker of the difference between Chinese and Western culture is Chairman Mao's calligraphy that adorns the Monument to the People's Heroes in Tiananmen Square. We cannot imagine a prime minister, even one with an extraordinary ego, encircling the Centennial Flame in front of the House of Commons with his own handwriting.

III

An unexpected thrill at the ink factory was the chance to touch the catalogues: the *Fang Shi Mo Pu* (the "Ink Manual of the Fang Family") and the *Cheng-Shih Mo-Yüan* (the "Ink Garden of the Cheng Family"). Mr. Zhou explained that these weren't the originals, of course, but facsimiles of the design books of the classic ink makers.

Fang issued his catalogue in 1588 (having worked on it for five years—somehow, presumably, without Cheng knowing) in eight volumes illustrated with over four hundred specimens. It was divided into two parts: one section of Fang's own poems and essays, along with complimentary articles from scholars and literati, and then a section of pictorial designs. Fang employed the best calligraphers and artists—and said nothing of Cheng. The laudatory essays also said nothing of Cheng. Cheng was outraged. Fang was a low-life he had taken in and nurtured, and now the depraved ingrate, having stolen the concubine, had succeeded in stealing the designs. Cheng gnashed his teeth and waited.

Six years later, around 1594, Cheng began to compose his own catalogue, which would be twice the size and with colour printing much more brilliant. It would have even more accolades from scholars. It would eclipse Fang in every way. The catalogue was a place for the settling of scores. In addition to the hundreds of articles and poems Cheng composed in jail, he persuaded over one hundred literati to contribute odes, prefaces, and articles for his work. Like the *Divine Comedy*, where Dante manages to get in his digs at his enemies and still produce enduring art, Cheng's catalogue proved transcendent. It was too late to be a commercial success, but that wasn't his primary intention. Where Fang had laid out his book to appeal to customers, Cheng arranged his designs in the manner of an encyclopedia. This was a reference book, something that not only promoted contemporary design but also consolidated the learning of the past. It's with Cheng that the title for people in the trade changes from "ink worker" to "ink maker." Indeed, in his catalogue Cheng refers to the gathering at the Orchid Pavilion, and the illustration of the Seven Sages of the Bamboo Grove is followed by an image of a literary gathering of Cheng and his friends: a clear and audacious bid to establish his status as a literatus—ink maker to, and of, the gentry.

Confirmation of Cheng's status comes from a curious source: the famous Jesuit priest, Matteo Ricci. He'd come to China in 1583 (coincidentally the year Fang began compiling his catalogue), and after being there for a year he concluded that it had three major religions—Buddhism, Taoism, and that of the Confucian literati. When Ricci first arrived he shaved his head and adopted the robes of Buddhist priests. After a time he realized that they had little prestige. He discarded the robes, and, "to gain greater status we do not walk along the streets on foot, but have ourselves carried in sedan chairs, on men's shoulders, as men of rank are accustomed to do.... Priests [are] considered so vile in China that we need this and other similar devices to show them that we are not priests as vile as their own." So, looking like a priest and walking like a priest was no good. Who did have respect? The literati.

On his initial encounters Ricci recognized their importance, but understood they believed in neither the afterlife nor the immortality of the soul. After he'd been in China longer he realized that many Confucians belonged to one of the other two sects along with their own. Being accepted by the literati would provide an in for his own teachings, for Christianity. In the summer of 1595 (Cheng's first summer in jail; it would be a decade before the two would meet) Ricci wrote to a friend in Macao that he and the other Jesuits had "adopted the special dress that the literati wear on their social visits." Ricci describes the dress in loving detail: it "is of purple silk, and the hem of the robe and the collar and the edges are bordered with a band of blue silk a little less than a palm wide; the same decoration is on the edges of the sleeves which hang open, rather in the style common in Venice." He strolled out (as far as his sedan chair) in purple and blue robes and embroidered silk shoes, stylish, flamboyant, powerful.

Himself an experienced printer, Ricci would have apprehended the skill of the *Cheng-Shih Mo-Yüan*. Meeting Cheng at a party in 1605, just after the catalogue had come out, Ricci offered him four Biblical illustrations for subsequent editions. This collaboration between Ricci and Cheng is significant for two reasons. It is through Cheng's inksticks, now transformed from a utilitarian product to a luxury good and a status symbol, that Ricci (who's been a player in official circles for over twenty years) seeks access to what he regards as the most important element in Chinese society, the literati. We could seek no greater proof of the power of the inkstick than the fact that the astute Ricci saw them as a means of insinuating Catholicism into Chinese life. Where people had to be persuaded into church to see the gorgeous multilingual Plantin Bible, Cheng's inksticks went out into the homes of the gentry and sat on the scholar's table, one of the Four Treasures of the Study, something no Bible could achieve. The first of the illustrations that Ricci gave Cheng was of Peter in the Sea of Galilee, being saved by Christ; he also contributed a brief essay on the print for Cheng's catalogue. As Wang Chi-chen notes in his *Metropolitan Museum Studies* article, there are mannerisms in the illustration that betray its Chinese copyist: the clouds, "which

The apostle Peter in the Sea of Galilee, from Cheng's catalogue,
combined Western and Chinese techniques. From Wang Chi-chen,
"Notes on Chinese Ink," Metropolitan Museum Studies *3.1 (Dec. 1930): 117;*
original in American Museum of Natural History.

are decidedly Chinese in feeling," and the face of the disciple, which "reminds one of traditional Chinese representations of patriarchs." But that probably worked to Ricci's advantage, since he wanted to stress the continuities between Confucianism and Christianity. The catalogue and the inksticks thus become part of a web of *guanxi* (influence) and power.

The collaboration was significant in another way, unforeseen by Ricci or Cheng. Chinese artists at the time never used groups of lines to give volume, and the surprise aroused by these images, particularly the Madonna, is recorded by many literati. One notes,

"The features are lifelike; the bodies, arms and hands seem to protrude tangibly from the picture. The concavities and convexities of the face are visually no different from a living person's." He puts the question to Ricci, who replies,

> Chinese painting depicts the lights (Yang) but not the shades (Yin). Therefore when you look at it, people's faces and bodies seem to be flat, without concavities and convexities. Painting in our country is executed with a combination of shades and lights.

The religious impact of these images is difficult to determine (the Chinese thought for a time that the Christian god was a woman, and were confused by a religion that declared there was only one god, yet offered a multiplicity of images), but the artistic impact was profound. As one scholar put it, "The *Ch'eng-Shih Mo-Yüan* records the very beginning of the Western foreign influence coming to China."

The pivotal points in history always prove to be less sharp-edged than we think—there were other printers besides Gutenberg, other ballpoint makers besides Bíró—but it seems clear that Cheng's inksticks were a crucial conduit for the artistic techniques of the European Renaissance. His catalogue, "The Ink Garden of Cheng," ink about ink, became more than an act of commercial rivalry and personal revenge, more than a passport to the highest social circles, more than compensation for a lost concubine and seven years in jail. The old ink maker had accomplished what the young poet had dreamed of: he had created a work of art that defined and extended the culture, that crossed borders, that endured through centuries, and that still guided the ink factory in his hometown.

IV

Leilei Chen, my Chinese research assistant in Canada, had asked, "Why are Westerners so interested in the Ming period? We do not find it so interesting. We prefer the Tang, or the Han." Part of it is that Chinese history is so vast, and the Tang dynasty (618–907) presents no easy parallels to the West—when Li Po was crafting his exquisite lyrics about inksticks in a brocade sack, my ancestors were sacking each other's huts and listening to how Beowulf slew Grendel. But the Ming dynasty (1368–1644) overlaps nicely with the European Renaissance. Fang and Cheng dovetail with Shakespeare: Fang's catalogue of inkstick designs appeared in 1594, a year before Shakespeare began writing *Titus Andronicus*, and Cheng's 1605 catalogue, with essays proclaiming himself a man more sinned against than sinning, coincided with Shakespeare's production of *King Lear*.

Yet I think the real reason we're attracted to the Ming period is that it provides all the elements for a Westerner's notion of the Eastern exotic—moonlit lakes, exquisite silks, beautiful boys, gorgeous concubines—and because we love those eras of lush decadence just before the bubble bursts: the Paris of Toulouse-Lautrec, the London of Oscar Wilde, the Jazz Age in New York. The final years of the Ming were like one long Roaring Twenties. While the Manchus were sharpening their swords on the northern border and preparing to do an end run around the Great Wall, the Ming were doing their equivalent of the Charleston.

The word that comes up again and again in studies of the Ming is *pleasure*. Jonathan Spence in *Return to Dragon Mountain: Memories of a Late Ming Man* tells of Zhang Dai, a wealthy young gentleman who devoted himself to his lantern collection, to playing the *qin* (a form of long zither), to cockfighting, writing poetry, and eating crab. The Crab-Eating Club met on certain afternoons in autumn when the crabs were at their best, "sweet and velvety" with "jade-cream" juices. When the winter brought a heavy snowfall, Zhang Dai sat out to observe the landscape and his own moods, fortified with the hot wine brought by a servant, while a concubine sang an aria and another accompanied her on the flute. The Ming dynasty, in place for two and a half centuries, had seventeen years left.

On a hunting trip Zhang Dai and his friends took five courtesans, "all dressed as archers, in red brocade trimmed with fox fur." He seems to have been particularly fascinated by the independent scholarly courtesans, who could play the *qin*, sing, recite poetry, paint, and beat a man in a drinking match. He wrote a poem (no doubt with the finest inks) to the singer-courtesan Wang Yuesheng, famed for her beauty and remoteness—"a solitary plum blossom under a cold moon, cool and distant." He wrote sketches of his relatives, such as the great-great-grandfather who came first in the Beijing exams, an honour that forever called to account his descendants—like the spoiled Uncle Yanke, who bought an expensive inkstone, split it trying to knock off a protuberance, and so smashed it all up, even the mahogany base, and threw it in the lake. The Ming had six years left.

When the Manchus invaded in 1644, Zhang Dai escaped to the mountains carrying a few bamboo baskets of possessions, his writing materials, and his manuscript of the history of the Ming. He was forced to abandon his library of thirty thousand *juan* to General Fang's troops, who consumed the books, "ripped them apart day by day, either to light their fires with, or to ... use them as wadding in their armour to ward off arrows."

The long party was over, Cheng and Fang dead some thirty years. Yet their inksticks have endured, and of course it would be the Ming period—indulgent, decadent, and glorious—that raised a utilitarian implement, the equivalent of a pencil lead, to an art form fit for an emperor.

*Bakong Scripture Printing Lamasery, Dégé, China (formerly Tibet),
with pilgrims circling the building. Photo courtesy of the author.*

From Chengdu to the Print Shop
at the Edge of the World

I

One of the pleasures of travel is backtracking. I returned to my hotel at the edge of the Nanjing University campus, weaving through the shoals of electric scooters moving up the bike lanes, past the food carts in the dip at the end of the block. Nanjing felt like home after the visit to Anhui.

I'd come to China not only for the ancient arts of fine ink but also for the modern promotion of bad ink: I hoped to pick up the trail of Milton Reynolds and his self-aggrandizing expedition to Amne Machin. In 1948 Reynolds and his scientific expedition had landed in Nanjing to coordinate with Dr. Sa Bendong of Academia Sinica. The Reynolds project was the first time Chinese and American scientists had joined together on an expedition in China. For Bendong, a graduate of Stanford, this represented an opportunity to develop a relationship with the American scientific community. However, it was not to be. He was bitterly disappointed that Reynolds turned the expedition into a stunt. Bendong felt that it soured American–Chinese relations, and Reynolds in his auto-biography portrayed Bendong as a fanatical Fu Manchu figure who predicted that China would rule the world. Two very different accounts, but I knew that Academia Sinica, the leading institution of advanced research in China, had articles about the expedition in their archive.

Yao Yuan met me for coffee at the New Magazine Café, a trendy place near the campus. I was exclaiming that this was the finest espresso in Nanjing when she said, "I must tell you about the archives. I didn't want to tell you before your visit to the ink factory. Apparently someone stole the earliest Chinese stock certificate in existence from an archive in Beijing. They have used this as an excuse to close the historical archives, not just in Beijing but all over the country." The Academia Sinica archives were closed indefinitely.

As an academic I was disappointed. As a writer I was relieved. By now, I'd found that I was less interested in Reynolds than in the terrain. Plus, I'd come across a reference to Dégé, site of the Bakong Scripture Printing Lamasery, "perhaps the heart of the Tibetan world" with over 270,000 engraved blocks of Tibetan scripture, representing some seventy percent of Tibet's literary heritage. And it was more than an archive; it was a working print shop: hundreds of workers hand-produced more than 2500 prints every day. The journey out there would put me on the northern route to Tibet, and my guide-book soared into Victorian adventure rhetoric at the prospect: "The disrepair of roads here exceeds those on the southern route and offers a real test of the mettle of any mortal who dares set upon them." I wanted to be one of those mortals, to brave the three days of eight-hour bus rides, a thousand kilometres from Chengdu to the edge of Tibet, on roads increasingly narrow, high, and dangerous. This would be more thrilling than trying to track Milton Reynolds.

From Beijing to Shanghai to Nanjing I'd been passed along from one friend of a friend to another, in the web of Chinese hospitality. Wonderful, but I felt as if my feet were never allowed to touch the ground. Used to travelling alone, I had begun to chafe. Now I'd be heading out with no one to meet me. I flew first to Chengdu. I wouldn't have time to visit the panda farm north of the city, but I would visit Du Fu's cottage. "It is very beautiful," said Yao Yuan. "A whole park in the centre of the city." Du Fu (712–770), the great Tang dynasty poet, is part of the national consciousness, so this seemed a fitting first stop on my journey to the west.

II

Chengdu, with its canals, winding river, and only fourteen million people, feels compact and charming after Beijing with its nineteen million and Shanghai with its twenty-three million. Jin Jiang, the "Brocade River," is the size of the Seine; it has lots of bridges, and its tree-lined banks are crowded with barbers, foot-callus filers, and potato-chip sellers working their trades.

Eventually I reached the park, where musicians and singers were playing from music tacked to trees and young children were running among the old men playing checkers and mah-jong. I wandered through the garden under grey skies. The shrivelled lotus flowers coming out of the mud—symbols of creativity—just looked sad, like crumpled paper from rough drafts. In the centre, at the replica of Du Fu's thatched cottage, I read the placards in Chinese and English that declared it was Du Fu's concern for the people that made him great. Then I read the translation of Du Fu's famous lines on the loss of his son, who starved to death in 755 CE:

> Wine and meat go bad in lordly mansions
> Bones of those dying in chilly wind are scattered on the road

The *China Daily* had been reporting on corruption in milk production, in bridge building, in the construction of the schools in Sichuan that had collapsed in the earthquake. Lives risked, lives lost, to make a few more bucks. *Bones of those dying in chilly wind are scattered on the road*. Probably most people in China could relate to the dismissive neglect conveyed in those ten characters. I had read the poem in other translations. Jonathan Waley renders it with outrage:

> Behind the rich man's gates, the obscene stink of feasting
> By the roadside the bones of people who froze to death

Rewi Alley rants and lets the rhythm go:

> There comes the reek of wines
> And meats that rot inside the gates
> Of those rich, the bones of the
> Starving and cold are strewn along
> The roadsides

David Young, aloof and spare, preserves the tight structure of the original:

> behind those red gates
> meat and wine are left to spoil
>
> outside lie the bones
> of people who starved and froze

Such latitude is possible because the poem in Chinese is always, in the Western sense, a skeleton. I would think different translations were different poems, and then, as when you encounter an old friend, unseen in years, who is unrecognizable before the bone structure underneath twigs the memory, I would see the core and know it as the same poem. Even David Young's version is an amplification. The translator is working with just ten characters, and all characters in Chinese are monosyllables:

朱门酒肉臭
路有冻死骨

red door wine meat spoil
road has cold dead bones

Ten syllables, as iconic in Chinese as "To be or not to be, / That is the question" is in English, yet sliding between the literal and the figurative to a degree unmatched in our language, and always elliptical. It helps to know that a red door signifies a rich man's house. No wonder diplomacy falters.

I moved on to a circular pavilion that displayed Du Fu's poems in various calligraphy styles. The poems were translated into English, but that was the least interesting aspect. Here the viewer encountered three or four versions of the same poem, on boards two metres high, in different scripts. Characters in the tight Regular script opened up in Running Hand and then took off in Grass Hand. Even though I couldn't read the characters, I had begun to recognize the scripts.

I'd read up on calligraphy in books by Chiang Yee and Tseng Yuho, and then found a good introduction to the characters of the five styles on the website of Montreal tattoo artist and calligrapher Ngan Siu-Mui. The ancient Seal script (which always looks kind of spidery to me because the strokes have no shading) looks more like a pictogram than words. Clerical script, which emerged during the Han dynasty, introduces variations in the thickness of the brushstrokes; it is "smooth and elegant." In Regular script, the most common calligraphy style, all the strokes are placed carefully and distinctly, the brush lifted from the page (this is what Lisa was trying to teach me). It is simple and neat, the most legible. With Running script, sometimes described as "semi-cursive," the strokes can run into each other. It's less angular, more rounded than Regular script. (It will give you a tattoo that is "stylish, spunky and charming. It shows audacity but does not exceed the limit.") Pushed to that limit of legibility, Grass script ("cursive") is even more rounded; the characters are simplified to a kind of shorthand, unreadable by most except the literati. Whole characters may be written without the brush leaving the page. Chiang Yee speaks of its uncurbed force and rapidity, and Ngan says "People call it 'crazy writing.'" Described as vivacious, daring, and dynamic, it can be almost frantic (a tattoo in this style will mark you as "wild and untamed"), but the style is revered for its abstract beauty.

A panel from a scroll by Dong Qichang, with the beginning of a poem by Wang Wei: "Looking Down in a Spring-rain on the Course from Fairy-mountain Palace to the Pavilion of Increase Harmonizing the Emperor's Poem." From the Mactaggart Art Collection, © 2014 University of Alberta Museums, University of Alberta, Edmonton, Canada.

In the pavilion I looked at three versions of Du Fu's "Welcome Rain on a Spring Night." The first was by Guo Moruo, who died in 1978, but who employed ancient styles. The second was by Deng Tuo, who was minister of publicity for the Central Communist Party and publisher of *People's Daily* (government officials were expected to be good calligraphers even as late as the 1960s), in Running and Grass hand. The third was by a woman, Huang Zhiquan, a scholar and member of the Sichuan Research Institute of Literature and History. She had lived to be eighty-five, dying in 1993, but her calligraphy seemed to me youthful, exuberant. I liked all three. From the samples in books, I'd thought I didn't like Regular script, preferred Running Hand or even Grass Hand, but now I was beginning to appreciate that calligraphy is like music: sometimes you're in the mood for Bach, sometimes for Miles Davis. Further on I saw a wild panel by Li Shan, a calligrapher who had settled in New York. I imagined him in the Village or in a loft in Soho, listening to jazz, hanging out with Jackson Pollock. I was beginning to get it: the poetry wasn't just reproduced by but embodied in the splash and swirl of the ink.

Later I would meet Walter Davis, who as a high school kid in Kansas fell in love with the calligraphy of Dong Qichang in a travelling exhibition of Chinese art. He went on to become a professor of Asian art, fluent in Chinese and Japanese, very serious, yet with the boyish enthusiasm that propelled him into his studies. Walter praised my untutored response to the Du Fu poems, then kindly but firmly told me I must distinguish between script type and personal style. He said that just as a singer working in different genres will retain her own manner of voicing, so a calligrapher will employ different script types, but that his distinctive personal style will shine through in each. "And don't forget social deployment," Walter continued. "Those inscriptions would have been chosen

both for their appealing calligraphy and the presumed fine character of the writers. The physical traces of one's brush were thought
to be indexical markers of one's moral fibre. Guo Moruo and Deng
Tuo are in very good standing with the CCP for their socialism."

So you're simultaneously reading the poet and the calligrapher.
Connoisseurs of calligraphy would reflect upon the calligrapher's
personal struggles and accomplishments as well as the writer's mental
state at the time of writing. This seemed a lot to take in, but then
I remembered the Billie Holiday recording of "Summertime," that
much-covered Gershwin song. Billie Holiday was only twenty-one
in 1936, and it was her first hit, yet today we hear the whole career
of pain and loss. She gives a bitter irony to the line "The livin' is
easy," stamping the tune as her own. Even for first-time listeners
there's something about her timbre and phrasing that raises the
hairs on the back of the neck. And so it must have been when the
young Walter Davis encountered the world's greatest calligrapher
in Kansas—intellectually baffling (he knew no Chinese), yet an
aesthetic revelation. Movement in ink that launched a career.

III

I caught a cab to the central square, photographed the statue
of Mao with his arm outstretched toward Starbucks, and then
headed to the luxurious Tibet Hotel: beautiful hostesses in traditional Tibetan dress, a restaurant for Tibetan specialties, a pillow
advertisement for Tibetan massage, and a gift shop filled with
Tibetan carvings and textiles. The cheapest scarves were five times
anything on the street. You enter the hotel's sumptuous lobby
through glass doors, but I had the uncomfortable feeling I was
behind a "red door."

The in-house travel agency specialized in tours into Tibet. How much, I asked, would a private driver cost to go to Dégé? The travel manager looked astonished. "It is very inconvenient. You would have to have a very good car."

"That's okay, I know buses go there."

"I will phone. But it is much better to go with a team. We have a tour going to Kangding. Nine hundred yuan for three days, including hotel."

"But I don't want three days. I just want to go to Kangding and keep going. It's not even halfway to Dégé."

"I will phone, but it is very difficult." The tour did tempt me. I didn't want to be stuffed onto a slow local bus, and my pronunciation was so poor I couldn't even communicate place names. I baffled the travel agent when I said "Kangding" (Kayng-DING). When I pointed to the map, she snapped, "KONding!," swallowing the second half so that it was almost "KONdng." I had reversed the accented syllables, distorted the vowel, and wasn't even close to the correct tone. I would have stood a better chance if I'd asked for "condom!"

She got off the phone. "It is very difficult. The road is very narrow. Snow will come soon. He say twelve thousand yuan." Twenty-four hundred dollars! For that I could buy an old jeep. The bus ticket to Dégé was 330 yuan. She repeated that travelling alone would be inconvenient and very dangerous. The tour would be very safe. Finally she played her trump card. "You are travelling alone. You know, there are Tibetans out there." I told her I'd met Tibetans when travelling in Sikkim and had found them very friendly. She looked dubious. I asked if she'd been to Dégé. "Oh, no!" she said. "No. I am from Chengdu."

I knew China was mounting a multi-pronged attack on Tibetan culture, eradicating and diluting, freezing some folkloric scraps for tourist shows. And here was this hotel, devoted to purveying the art and culture of Tibet, with a travel agent who shuddered at the thought of running into a real live Tibetan. I wanted no part of her tour. I got the desk clerk to write out *Kangding* in Chinese characters on the back of a business card. The other side said, "Take me to the Tibet Hotel." If I couldn't negotiate a bus ticket I'd come back and visit the panda farm.

IV

The bus charged up steep green valleys, wrinkled like a Shar Pei's forehead, luxuriant with bright orange persimmons, rickety footbridges across roaring creeks, bright vegetable gardens cleared in the forest. The two young guys behind me smoked constantly. Back home I'd lamented that blues bars just weren't the same now that no one smokes, but three hours in a cloud of cheap tobacco fumes cured me of all nostalgia. In Kangding I took a room in a little hotel near the bus station and washed the smoke out of my hair. After a dinner of spicy chicken I burrowed in under the covers, grateful for the cold after the stifling hotels in Nanjing and Chengdu. I'd come from a lush lowland at 500 metres to a stark landscape at 2600 metres, and it felt like another country. I fell asleep looking at the fiery Tibetan symbols on the ceiling. Later I would learn that it *was* another country: large chunks of the former Tibet had been bitten off and ingested by Chinese provinces, and before 1959 Kangding had been the capital of a Tibetan kingdom.

The next morning I made my way down the black, unlit street to the bus station. By the time the sun came up we'd left the forest and moved out onto the vast Tagong Grasslands. The guidebook refers to the "chocolate drop" hills, but they look more like great slouching beasts. There are no trees to give scale, and after a while you notice that the hills have clumps of black dots on their steep sides, with tiny bright specks beside them. The clumps turn out to be herds of yak and the specks the women tending them, and you realize your sense of proportion is completely skewed. These are hills become mountains, and the plains abutting them are vast. I dozed and woke and still they were there, the monstrous hills. The road signs had Chinese characters—they used highway-sign style, a Regular script with no modulation, assertive and threatening—above the sweeping Tibetan script, but we were only technically in Sichuan province.

The windows clattered in their channels as the driver banged through the potholes. I learned to "post," to lean forward in the seat and get my spine straight. If you were licking your lips as the bus smashed over a bump you'd bite your tongue off. This must have been what it was like to travel by stagecoach. I sat with my pack on my lap the whole eight hours, hands resting on top just below my chin. I had smokers behind and across from me, but at least my seatmate didn't smoke. The porous-weave curtains at the window were crisp like old hockey socks and smelled of stale sweat. When the bus rounded a sharp corner the cloth would lurch out and touch my cheek and gently stick. It was my second day without a shower, and there was no hot water that night in Luhuo.

On day three we left the plains, gassed up mid-morning at Ganzi—a town that, at 3400 metres, looked out to spectacular snow-slung spires against a deep, blue-black sky you only get at high altitudes—and stopped for lunch in Manigango. A man jogged by with a bleeding yak head on his shoulders. His friends bungeed chunks of the freshly butchered yak onto a motorcycle, piling them up until the front wheel lifted, leaving nowhere to sit except on the bloody meat. More slabs lay on the road with the blood-soaked pelt. The Chengdu travel agent would not like this.

The road wound along a river, shallow and clear, with the peaks rising beyond. In one of those weird through-the-looking-glass moments that happen when you travel, it looked exactly like Highway 22 south of Calgary, but we were already higher than the highest of the Rockies, approaching the 6000-metre Chola Pass. The guidebook had said, "It is not uncommon for buses to overturn on the icy, hairpin roads." What precisely constituted "not uncommon"? Did that mean twice a year? Once a week? Just what were our odds?

And "overturn." Did that mean the bus slipped like a stout man on a banana peel and went over on its side and lay there in the middle of the road? Or did it mean it slid off into the meat-grinder abyss, turning over and over, gored, smashed, ripped, crumpled, bodies tossed, luggage tumbling, coming to rest at the bottom of the gorge with blood pouring out?

Best not to think about it.

The road turned to a single dirt track. The drivetrain whined, the windows rattled, clouds swirled behind the peaks. The tiny crags along the ridge looked like razor wire or broken glass embedded in a guard wall. At the top of the pass the Buddhists in the bus surged to the windows and threw coloured paper charms with blessings written on them out into the air, shouting "Kiki Soso La gyal lo!" ("Victory to the Gods!"—apparently the evil gods fight the good gods at the top of mountain passes; I was glad that the Protectors of Rickety Buses had prevailed.) The charms littered the landscape. This was why Buddhist print shops thrived—these charms were like confetti and every Buddhist would have a bundle.

Like prayer flags, the charms produced positive karma and purified negative karma. You throw the *lungta* papers (*lung* means "wind," *ta* means "horse") only from high places, so that they'll fly everywhere with their positive energy. They're never thrown in the city or public areas because stepping on a charm brings negative energy. Nor are they thrown in lakes or streams; the devout press engraved metal plates onto the water and the blessings are carried away.

Above the swatches of *lungta* on the ground, lines of prayer flags snapped in the wind. The flags are always placed in sets of five, the colours representing the elements; they're arranged from left to right in order: blue (sky), white (wind), red (fire), green (water), yellow (earth). Traditionally they're hung in high places, not so that they can carry prayers to the gods—which is what I thought—but so that they can spread compassion out into the air. These ones at the top of the pass were ragged, shredded by the wind whipping through them, but they were never meant to be permanent. As the ink fades, the prayers pervade the universe; then the individual flags are replaced, new life succeeding old in the great cycle.

What inks would they have used to transmit these prayers? Probably a variety, depending on the location. Tibetan ink recipes were similar to the Chinese, but meatier, and always with one eye on winter. You can even make an ink with "animal sinew free from blood, nerves, and integument" that works particularly well on clay—but it congeals if there's too much sinew. As usual, local ingredients figure: an ink can be made quickly from radish, though it fades (excellent for grocery lists, I would imagine), a dark-red ink from the juice of the ripe fruit of the *ricinus communis*—the "castor oil plant," and a black ink from dung and from puffballs, both of which are put into the fire until glowing red, then thrown in water, and finally dried and ground.

For the best Tibetan inks the basic component is still soot, which you get from the top of the flame of an oily butter lamp (add a little goat's blood to enhance the binding), or from burning larch or birch bark. Then you add a solution of animal-hide glue, which has cooled and congealed to the point where your tongue can no longer pierce it (though even the translator of the nineteenth-century handbook I'd read admitted that no one would want to touch it with their tongue). Press this in a leather bag and add a pepper decoction to keep it from freezing in winter. One can also add the "clean lymphatic liquid of weak animals" to make the ink shiny and keep it from blotting in wintertime. The problem of freezing was a recurrent theme. I knew that priests on the Canadian prairies in winter had to thaw their ink, and of course it would be a factor on the Tibetan plateau, but I hadn't considered the consequences of ink freezing—and would wish I had.

The bus travelled down past the shimmering Xinhua Glacier, then into a warm valley full of pine trees, and then into a deep canyon, rocks closing over the sky above. You could imagine soldiers unleashing arrows and boulders on the single-file invaders below. Though the landscape opens up again and becomes more welcoming for a time, Dégé itself is in a steep narrow valley, with a shallow river rushing through the middle of town. On the outskirts we passed a man hewing logs for his house, three women with hand tools digging a foundation out of the rock, and an old man bent almost in a right angle carrying a bag of rice on his back. From behind he looked like a bag of rice with legs. At the dank market by the river men butchered yak on the curb. Yet in the centre, pedestrians in North Face jackets strolled along a tiled mall where golden prayer drums faced pretty-girl advertisements for cell phones. Across the river the commercial street sported new lampposts—that universal sign of civic upgrade—entwined with gold dragons. On the steep slope leading up from the river, modern apartment buildings pushed back wooden houses (built in a brown half-log style uncannily like 1940s Canadian National Park buildings), yet even these had bright paint unweathered by a single season, and many had unpainted additions, the wood fresh and tawny. Dégé was booming. Money from Beijing?

Dégé means "peaceful land" in Tibetan. It was once the capital of a kingdom; at 3270 metres, it's more than 500 metres higher than Kangding, and was as well protected as Shangri-La. But its isolation couldn't protect it from the Cultural Revolution. Chairman Mao had decreed that the new society could emerge only with the destruction of the Four Old Things—old ideas, old culture, old customs, and old habits—and Tibet represented everything old. Dégé's print shop and hospital were denounced and marked for destruction. In the winter of 1967 Tibetan and Chinese Red Guards broke into the temple, intending to burn the printing blocks for heating and cooking, but the secretary of the County Committee phoned Premier Zhou Enlai, who personally intervened to keep the blocks locked away. I wondered if they'd last another two hundred years.

Tibetan calligraphy was introduced in the seventh century, derived from India but influenced by China, and although calligraphers used a reed pen instead of a brush, the inks were similar. While not an ancient tradition, calligraphy is important to Tibetan identity. In an interview with the Xinhua news agency, Kalnor, a Tibetan calligrapher, recalls how he used to use a bamboo stick, its sharpened end dipped in soya sauce, to write on a *jangshing*—a fifty-by-twenty-five-centimetre birch board that children would practise on for two years before being allowed to use paper. "When we wrote on the jangshing, we sat on the floor with legs crossed and wrote every stroke with patience," he said. "It was not just a practice of calligraphy—it was also a process of extreme concentration and meditation." Now, with exercise books and computers replacing the *jangshing*, Kalnor has been travelling across Tibet collecting them; he feels they're important to the Tibetan collective memory. Several Tibetan scripts, including one unique to the Dégé region, have been added to the Chinese national list of intangible heritage for conservation. But many artists outside Tibet are concerned about the erosion of Tibetan identity. Phuntsok Tsering, a Tibetan calligrapher living in Germany, has created contemporary art out of the older traditions, and uses his calligraphy to honour the self-immolation of Tibetan protesters.

V

After checking into the Golden Yak hotel by the Dégé bus station, I set off to find the lamasery. I found a stream of people striding briskly from the back of the building—it was after five p.m., so a tide of workers going home, no doubt—and I imagined the bustling print shop within. I'd come back the next day when it was open.

I'd learned from William Milne, a Scottish Protestant missionary writing in 1820, that the lamasery's print blocks were made from the pear or jujube tree: "they are of a fine grain, hard, oily, and shining; of a sourish taste; and what vermin do not soon touch, hence used in printing." They smooth the block with a plane and then rub the surface with rice paste, which softens and moistens the board so that it more easily receives the impression of the character. The transcriber writes on the block in black ink and then sends it to the block cutter. There the block is covered with oil to make the characters more vivid, and the carving begins.

In some shops, the hard work (horizontal strokes) was done by accomplished carvers and the easier work (vertical strokes) by apprentices, so there might be as many as four workers to carve the text of a single woodblock. The goal was to cut rapidly and cheaply, not to reproduce the expressive qualities of the original calligraphy.

Simon Winchester in his account of travelling up the Yangtze, *The River at the Centre of the World* (1996), describes the scene at the Dégé print shop. Two boys "would sit facing each other, the one holding the block between his knees with the lower end resting on the floor. His partner would then swiftly roll an ink wheel down over the block ... and then the first boy would with equal swiftness take from a pile on his right a sheet of fine mulberry paper, about thirty inches by five, and place it on top of the ink-glistening wood." I liked that phrase "ink-glistening wood" and was looking forward to seeing the shining blocks.

Block printing would seem a cumbersome method compared with movable type (which was available to the Chinese—it was invented by one Bi Sheng in the 1040s, four centuries before Gutenberg), but unlike English with its mere twenty-six letters, the Chinese language has thousands of characters. There were other advantages to block printing. Printers didn't need a foundry for casting type, expensive machines for printing and binding, or premises with costly rent for housing equipment and products; they had only to house the woodblocks. Carvers needed only a handful of tools along with a small table and a stool, furniture that could be located anywhere, from the corner of a bedroom or court-yard to the deck of a ship. And unlike typesetters in the West, block cutters (many of whom were women) didn't have to be able to read. Wages were low, and a good cutter could cut 150 characters a day. Thus overhead was minimal. Plus, a single block could make ten thousand to forty thousand impressions—maybe not enough for the run of a Stephen King novel, but still plenty—and often the blocks could be recarved. As for the printing, two workers could run off two thousand copies per day, the equivalent of an English print shop's production in the early 1800s.

AT THE GOLDEN YAK HOTEL (no heat, no hot water) I wanted to send triumphant emails telling everyone I had arrived, but Dégé had no internet café, only a single public computer in the back of a cell-phone shop. Buddhist monks in their burgundy robes crowded around the counter discussing the different models. A strange meeting of old and new, I thought (the tourist's cliché), yet maybe not so odd. They'd been spinning words off prayer wheels and flinging them into the ether for centuries—was that so different from texting? I crossed the road to a little restaurant and ordered a bowl of noodles, using the technique from a guidebook: look at what everybody else is having, and when you see something you like, say *"Wo yao zhi ge!"* ("Whoa yow chu guh!"—"I want that!"). When they say a bunch of things back to you that you can't understand, simply exclaim, *"Dui!"* ("Dway!"—"Right!"), unless you see something really revolting, and then you reply, *"Buo yao!"* ("Boo yow!"—"Don't want!"). You do have to say everything as if it has an exclamation point after it or they don't understand; otherwise you'll sound like a drunk from Alabama would to a Canadian—"Aaaah wooont theyyyattt!" Sounds stretched to incoherence.

I was cold all the time and had a faint roaring in my head like a far-off waterfall. I hadn't spoken to anyone in three days. My only conversations were with my coil notebook, the flow of ink my consolation. I was reading *Red Dust: A Path Through China*, by Ma Jian, a Beijing writer who sets out to discover his country. He too takes refuge in his fountain pen. Intimidated by the climbers below Everest, he writes, "I heard expedition teams drive past here on their way to Base Camp. I take out my fountain pen and mutter, 'I'll just fill up my pen if you don't mind.' ... When the cartridge is full I discover the ink is blue, not black as I had thought. I hate writing with blue ink. A wave of tiredness sweeps over me." (Anyone who uses a pen knows that sense of dismay upon finding that you have the wrong colour—a blue too light, a black too grey—and knowing you'll have to mentally work against it as you write, that it somehow makes your thoughts inauthentic. A bright blue can make sombre reflections look whiny. You want to write in the margin, "This isn't really me, imagine this in jet black!")

Ma Jian was also, like me, feeling the altitude: "I understand how hard it is for man to live in the sky. The air is so thin I have to breathe through my mouth.... Tibet's high plateau is no place for the Han." He likes the Tibetans: "They are kind people, and when they accept you as a friend they will trust you with their lives. They are not as crafty or sly as the Han." And he feels they have a right to be angry: "Imagine if you invited some friends for supper and they decided to move in and take over your house." This too I identified with. And the title, *Red Dust*, refers to the mist of worldly illusion that we must strive to overcome. I was about to discover my own red dust.

IN THE MORNING I walked up to the big double doors of the lamasery, following some Tibetans through the defile of the inner courtyard and then into what I assumed would be the print shop. Everything was dark. This did not bode well. Once my eyes had adjusted I saw that we were in a big room with scuffed wooden floors and red pillars, scraped at the bottom as if they'd been gnawed by beavers and decorated at the top with designs in blue and bright red with gold scrollwork beneath. Against them, piled higher than a man's head, were wooden stools stacked and tied. This was the print shop and these were the stools on which the boys would sit facing each other, inking the carved surfaces, pressing the paper. Later, back home, Leilei, my Chinese researcher, would explain: "They put water on the blocks to moisten them. But when it gets too cold the water freezes, so they cannot keep printing." I had succumbed to the illusion that this would be the Chapter That Writes Itself. I should have known better—the gods of writing always punished my presumption. The print shop had probably closed for the winter before I arrived in China. Those workers striding from the lamasery yesterday? Not workers: pilgrims. Making their thousand circumambulations that amass good karma for the body, the mind, and the soul.

Snow was falling now, and I wondered whether the pass would be closed, whether the bus had slid into the icy abyss. I made my way over the slick stones at the gates to the pilgrims circling the building. They do it very purposefully, at a very brisk pace, which I guess you have to if you're doing it a thousand times. I stood watching, and then, having nowhere else to go, I joined them.

*A scholar's desk often included a stone for contemplation, along
with the "Four Treasures of the Study": paper, brushes, inkstone, and ink.
© Lao Hu Kaiwen ink factory, courtesy of the author.*

Home, and Three Calligraphers

I

China was a mystery. And yet, as I read into it, I found hints for my own domestic situation. Hongwu wanted to "immobilize the realm." Merchants were suspect precisely because, unlike farmers, they travelled. The first Ming emperor believed that twenty *li* (twelve kilometres) was the farthest anyone should go, and he decreed that to travel more than a hundred *li* (fifty-eight kilometres) you needed a certificate. Without one you'd be flogged eighty strokes. And if you dared to travel abroad without official documents you'd be executed when you returned. Social mobility was similarly fixed: if you were born into a family of artisans, you remained one. You knew your place, geographically, socially, economically. That fundamental Daoist text, the *Daode Jing* (*The Book of the Way*), set forth the ideal back in the sixth century BCE: "Though adjoining states be within sight of one another and cocks crowing and dogs barking in one be heard in the next, yet the people of one state will grow old and die without having had any dealings with those of another."

This reminded me of Hsing's mother, who kept the alarm system on the house activated at all times—not so that she could call the police if brigands broke in, but so that she could monitor her children's comings and goings. The question was less "Where are you going?" than "Why are you going?" The subtext: You have family. What need is there for you to go out? Chinese mothers frown upon anything that could be regarded as *xian guang*, and that included moving around: in fact, "fooling around," *xian guang*, is composed of the characters for *idle* and *to stroll, go about*.

So of course in North America, Chinese children take off. But they do it in their thirties, not in their teens. And though they may travel halfway around the world, they return not far from home. "YOU WANT WATERMELON?" says the voice on the answering machine, five times louder than any other message. "WHY YOU NOT PHONE ME?" You think you can change this? Look to the emperor Hongwu. Any Chinese mother would feel that fifty-eight kilometres was more than enough to travel unsupervised, that twenty *li* was actually the ideal; you can go shopping and come back. Any Chinese father would concur that foreign travel without documentation was deserving of execution.

The first road trip I made with Hsing was to an academic conference in Irvine, California. We were going to camp our way down the coast, make a holiday of it. I still hadn't met her parents. The day before we went she sighed and said, "I'd better tell my parents."

"What? You haven't told them yet? Why not?"

"They'll be set against it. They won't approve."

"Why didn't you tell them a long time ago? Then they'd have had time to get used to the idea."

"You don't understand. It would just give them more time to chastise me. It wouldn't change anything."

The next day I pulled up in front of the house (in an unfortu-nately phallic red Ford Probe) and popped the hatchback. I walked up to the door. A compact Chinese man was vacuuming the stone hallway. "Hi," I said, "I'm Ted," and I stuck out my hand. He ignored it.

"YOU AH TOO OLE TO DATE MY DAUGHTAH!"

"Pardon?" I said. His accent was very thick.

"TOO OLE! TOO OLE TO DATE MY DAUGHTAH!!"

Ah. Too old. I put my hand down. He continued to vacuum the stones savagely, which I now saw were immaculate and had been for some time. What he wanted to be doing, of course, instead of clattering the power head across the slate, was vacuuming my face off.

"WHY YOU GO CALIFONIA?"

"Um, it's a James Joyce conference."

"HAH."

Which pretty much ended the conversation. I walked back to the Probe. Hsing came out with her bag. "Well, that's my dad," she said with a smile, as if to say, "Such a kidder."

"It's not you," she said helpfully. "It's just that you're white. And my mother said 'If you're going to go out with an old guy you should at least pick a rich one.'" It was five years before I was invited to a family dinner. I was assured that this was normal for an aging white guy, and the fact that they hadn't actually forbidden her to see me was proof of enormous good will.

Years later, when I crashed Hsing's motorcycle and wound up in the hospital for a prolonged period, it was Fang, her father, who came every day and talked to me in his thick accent that I now could understand. He came more often than anyone else. "Why does he do this?" I asked Hsing.

"You're family," she said, surprised at my inability to grasp the

obvious. Family. Source of all strife, yet the basic unit of civilization. It would be five years more before I discovered, through his
calligraphy, another side to this severe patriarch.

After my China trip I organized the first Christmas dinner
with Hsing's family and mine: a feast of turkey and fried tofu,
mashed potatoes and sticky rice, carrots, noodles with chicken,
spicy spinach, pork dumplings, pumpkin pie, red bean cakes, red
bean soup, and an English Christmas pudding with hard sauce. At
these family dinners where usually no one knows what to say to me,
Hsing's Aunt Millie and Uncle Tak happily talked about ink.

"There are some inksticks that cost a thousand dollars, and
you mix them with wine," said Uncle Tak. "If you mix them with
water they smell terrible but if you mix them with wine they smell
wonderful, and they make a thick black ink that will never fade."

I asked about the problem of the young not learning calligraphy anymore. "It is the way it is with historical things," said Aunt
Millie. "There is just no time to do these old things; you have to
make choices."

"Yes, but they still learn in school, and so after university some
will take it up again," said Uncle Tak.

That day I'd been working on my character for *word*, six
strokes, top to bottom, left to right:

Lisa had made me a model to follow, slowly and precisely
forming an elegant character. Then she moved over to help one of
the youngsters with their *heng* strokes. Fang watched me.

"No. Too long," he said, meaning that my lower horizontal

stroke was longer than the top one. Forty years in Canada and he still barks out heavily accented monosyllables. As though he's training a dog, I used to think ("Sit!" "Stay!"), and resented it. But now, having been to China, having struggled to make myself understood with "Kangding!," I recognized that he was just speaking English with Chinese tones.

"All must be under," he said, meaning, I guessed, that the whole character must be under the tent formed by the top stroke. I tried again, and then he reached over, took the brush, and whipped off the character.

"See? Don't stop. All one." In his, the second horizontal line that angles back didn't stop but led into a swooping downstroke. Lisa had stopped and lifted her brush at that point. She'd also doubled back on her horizontal strokes, as with the *heng*.

Fang said, "No. Don't go back. One way only." I was shocked. Not stop? Not take the breath?

He did it again. Swift, stylized, slanted. Running script? The ink flowed, the characters appeared in a line, and I thought of a young man in a zoot suit. Cool. Raffish. The character of his character. Not what I expected from a taciturn chemist.

My nieces and nephew took turns, kneeling on the rug in front of the big sheets of rice paper, with Lisa and Fang hovering over them. My brother Norm, rock drummer turned lawyer, produced a particularly tasty *nah* stroke, the one that looks like a hockey stick, and earned applause.

There was a good feeling in the room; my plan to unite East and West through ink had worked.

II

The irritating thing about travel research is that you often come home and find what you need on your doorstep, and not newly arrived. As I was waiting to see my acupuncturist I was thinking about the failures of my China trip, not just Dégé but the fact that I hadn't managed to talk to a practising calligrapher. I'd started going to Dr. Steven Aung a decade ago after a motorcycle accident and now saw him a couple of times a year for a "tune-up," as he calls it, aligning my *qi*, tweaking the meridians. I didn't understand it, but I knew it worked. Initially skeptical, I'd gone as a guinea pig for an acupuncture class Hsing and other medical practitioners were taking with Dr. Aung. For months after the accident my neck was so stiff that I drove like the cartoon character Mr. Magoo—when I wanted to change lanes I'd put on my signal light and inch over toward the new lane, waiting for a honk or the sound of scraping metal.

That first time I went, Dr. Aung stuck a needle in my forearm. It felt like a mosquito bite.

"Oh!"

"What do you feel?"

"It's weird, it's like a coal chute opened up in the right side of my chest."

"Ah." He stuck in another needle. "What do you feel?"

"Nothing."

"Turn your head."

"That's amazing! I can see way to the side!"

"No good. Need more." Dr. Aung stuck in another needle.

"Wow! One more needle and my head will go all around like in *The Exorcist*." The medical students were delighted. The Skeptic had been humbled, the Master vindicated.

I'd heard that he did calligraphy, and sold pieces at charity auctions. Once we started my session I asked him about it. "Oh yes," he said, dropping a needle into the soft pass between my big toe and my second toe, "I do calligraphy every day. I become a different person when I do calligraphy." He put in another needle, tapped it, and I felt the deep ache of it lighting up the meridian. It felt like something surging, coming up against a blockage and then bursting through, like Drano dissolving the grease buildup and hair clumps. My eyes fluttered shut on their own and I drifted into the acu-zone.

Later in his office, where two aquariums with carp gurgled beside us, he said, "There are three kinds of calligraphers: academics, professionals, and one who is just interested and practises." He, obviously, was the last. "What I do is half calligraphy, half painting. Writing very fast, your energy moves, you are controlling your brain but not your hand." He wore brown pants and a sports shirt—he had taken off the white labcoat that marked him as the doctor. "Your mind has to sink in. The realization of yourself becomes the writing. Calligrapher-calligraphy. Even my handwriting changes. My grandfather's handwriting, now I can write it."

He spoke with a soft Burmese accent, the words emerging out of a murmur; he would often drop his articles and prepositions: "the practice of calligraphy" became "practice calligraphy," which blurred the distinction between noun and verb, enhanced the sense of flow, kept the listener grasping at possibilities. Which one? Both? With the fish tanks behind me I was hearing him through water.

I asked if the state he achieved with calligraphy was the same as when he practised acupuncture. "Oh yes! The same thing. You are in a deep somatic state. That's why the art of healing is there and the art of calligraphy is too. Because you heal yourself as you heal other people. But that's a different thing."

Indeed. I was getting lost, but I didn't want to backtrack. I knew that every day he did *qi gong* (a spiritual practice of meditation and movement that is also training for medicine and the martial arts). "Oh yes. Calligraphy is a *qi gong*. Because if you are only writing calligraphy for beauty, your strength is there but *qi* is not there. With every breath the pen is moving, but every time you breathe out you have to stop. Just like *qi gong* exercise. And also when you look at calligraphy you know *exactly* where he stopped his breathing. And that stop point you can tell how much energy has this man. If you stop with only tiny little line then your *qi* is very short. It's like a figure skater. The movement is continuous, and though sometimes you jump up, when you land, you must land with force."

I think I see, but I'm thinking of the Bic pen ads with pretty girls in short skirts on skates.

"Practising calligraphy is not just practising writing. A lot of people make that mistake. Actually it is the practice of letting it go. People have lots of stress in their hearts; they don't know how to let it go. They have to use calligraphy to let it go."

My calligraphy attempts just induced stress in the heart, but I did not say so.

"Everything in our life, or happening, is related with each other. Every time I am writing my calligraphy my body is … free." He laughed. "Like a tree, you know when you see them blowing? Same thing."

I spoke of the willow trees around Xuanwu Lake in Nanjing, how blowing in the wind they looked like brushes, in fact like brush *strokes*.

"Ah. The willow tree is a negative tree, negative energy. You might wonder why Chinese love these trees? You love suffering … when you are in the higher level you never learn anything. When you are in the lower level you learn a lot. So with willow trees, you see all the willow branches are dripping down, dripping down, and when the wind blows they grow together like a brush. The Chinese enjoy that, and they look at the nature of the wind that blows. The tree then is movement. It's a stage of art, actually movement." He pointed out the window. "Look at all these trees, they are pointing heavenward"—he held his arms above his head—"they cannot write. Everything is all underneath, the writing. That is why the willow tree plays a very important part in calligraphy."

Hmm. I wasn't sure I got the willow tree. I knew that they were favoured by poets and scholars who would walk among them for inspiration, and that they were associated with compassion.

He'd mentioned that as a calligrapher he liked to know about the materials he was using, so I brought him back to that. I'd read of an ink maker whose ink would last one hundred years.

"Oh, that's true. The cheap one may last you a certain amount of time, but a good one like this will last you five years. I have the ink of my grandfather." He told me how when he was a child he and

A basic inkstone is not much bigger than the palm of a hand,
but ornamental inkstones like this one can be more than a metre long.
Photo courtesy of the author.

his two brothers had assigned days to grind ink for their grandfather. His days were Monday and Wednesday. The grandfather always said to grind one hundred times, but they would try to run away after grinding only ten times. Still, he learned to appreciate doing it the right way. "When you do hundred times the ink and the water they mix. And it looks *so* beautiful. And the smell all comes up."

I told him that my ink was sort of grey and a bit blotchy. "Ah, you are not good enough yet. You have to keep grinding."

I asked if he was doing any calligraphy at this stage, and he said no, the first step was grinding for the master. "When you are grinding, the master looks at you in your corner to see if there is good enough concentration. If you are grinding too fast the ink doesn't come out right, and if too slow no good. And always clockwise. When you do it counter-clockwise the ink comes out different. It is the same thing in acupuncture; it makes a difference which way you turn the needle."

Only when you have the feeling for grinding the ink are you given a brush. "This guy who worked for Confucius for the first ten years, he was grinding ink. My grandfather always said, 'You are not good enough yet.'"

I saw now that in our calligraphy lessons Lisa was in fact being polite.

"You see," said Dr. Aung, "grinding ink is the art of your doing." Again the burble from the fish tank made his meaning unclear. Did he mean "the heart of what you're doing"? Or was it the same thing? I had the sense that he was trying to separate concepts for me that for him weren't separate. Grinding ink *was* calligraphy.

I told him about my visit to the Anhui ink factory, and mentioned my unproductive trip to Dégé. "Ah, you should have told me! I have a friend with an ink factory. It is fascinating. The people who make ink are not ordinary people—so many hundred years of writing and calligraphy. Also Tibet. I have been there many times. Not Dégé but Lhasa and other monasteries. You should have come on our trip; we visited many places others cannot go."

My loss. I should have prepared more fully. I felt like a ten-year-old who'd scampered off after grinding the inkstick only ten times. Yet I was glad that I'd made my way alone, and the trip to Dégé had been an excursion in humility if not a field trip in ink. Perhaps that had been the sign of the willow trees in Nanjing.

III

I'd registered for a lecture and workshop the next day with Arnold Chang, a well-known painter and calligrapher. Where Aung is round and slow, Chang is hard and quick. He's Asian but speaks with the cadences of a born New Yorker, and he has the New Yorker's scorn for pretension. As he began to paint, one of the artists in the group asked him what kind of ink he uses. "I use the cheap stuff," he said. "Whatever works. I know people who go on and on about brushes and inkstones and they can't paint a lick." Someone else asked how much to grind the ink. Chang was scrubbing it back and forth like an eraser on the inkstone. "Till it's dark enough," he said. He scoffed at the notion of grinding the ink to put yourself in the right mood ("organizing your *qi*!") and said he didn't like bottled ink just because it was too dark, but otherwise he would use it. I was surprised. Walter Davis, the professor of Chinese art, had told me that Chang's teacher had been C.C. Wang, "so he can legitimately claim to be one of the last of the true traditionalists."

After the lecture and the demonstration it was our turn. Chang set us up at tables with paper, ink, brushes, and inkstones. "What I want you to do is make spirals," he said, and warned us that it was harder than it sounded. "The idea is to keep the line round, like a hose, as opposed to a ribbon, which is flat. You know how when you're laying down a ribbon and you change direction you get a flat fold? Well, you don't want that. You want it to stay round, no kinks in the hose. So that water could flow through it." He added as a throwaway, "The old painters used to talk about *qi* flowing through the line."

We began our spirals. I reached the top of the arc and my line immediately kinked. "You can't push, you have to draw it out, lead it, and you don't use the wrist, you use the whole arm." My spiral looked like an old nickel or a stop sign—flat spots and angles rather than a smooth curve. This was hard, but I could now see, dramatically, the difference in the line. Then we were allowed to copy some basic lines of mountains. Huangshan, here I come! Well. Up and down, flat-ribbon turns at the top, kinked hose, no *qi* flow here. My neighbours were good, which I resented, though I was pleased to see they were having difficulty as well. I reminded myself that it wasn't a competition, that I'd come only for the experience. My art career had stopped in grade three when I was labouring to draw a horse and the teacher looked over my shoulder and said brightly, "Oh Ted, what a nice duck!" Every ten years or so I try to draw something, and the result only confirms my early lack of promise. The fact that I was in this workshop at all was a personal triumph.

Chang came around and offered advice to my tablemates (my hacked upside-down *W*'s were clearly beyond help). "The stroke comes from the arm. In fact it really originates here"—he touched the centre of his torso—"and the idea is that the energy, the *qi*, travels up and through the shoulder, to the arm, and through the brush." He was starting to sound less like the opposite of Dr. Aung; in spite of himself he was talking about *qi*. *The Mustard Seed Garden Manual* says, "Rocks without *qi* are dead rocks."

On the way out of the gallery I looked at one of Mr. Chang's landscapes again, at the way the mountains seemed to surge up from the river, and thought I was beginning to understand. Something was happening there, at the level of the individual line, that was not happening in our student attempts. It was a sense of vitality, not just the obvious energy like the movement in willow trees, but of life in things we had thought inert. I used to joke to Hsing that acupuncture was better than being stoned; it didn't happen every time, but when it did (whatever "it" was) I drove home slowly, conscious of the molecules in the air, fascinated by lawns, lamp-posts, sky. It was even better on a bike; I could feel each ripple in the road, sense the wheel turning. I couldn't catch it in words, but master Chang had caught that life of things in ink, in a little water and solidified smoke.

PART III

THE *Spirit* OF

Say: "If the ocean turned to ink
for writing down the colloquy of my Lord,
the ocean itself would be exhausted
ere the words (and wonders) of my Lord come to end ..."

—QUR'AN, THE CAVE 18:109

Ornate Qur'an with illumination in gold, Topkapi Palace Museum, Istanbul.
Photo courtesy of the author.

The Samarkand Codex

I

We changed planes in Moscow at midnight, and as we headed south into the blackness I turned to my brother Lloyd. "So, where exactly is Tashkent?"

"It's difficult to say," he said.

I'D COME BACK from China in the fall sated with detail, weary of travel. As the temperature dropped to minus thirty I read *The Splendor of Islamic Calligraphy*, tracked obscure articles on gallnut ink, and settled in with Orhan Pamuk's *My Name Is Red*, an Istanbul murder mystery, love story, and homage to the Persian painter-calligrapher Bihzad. I dreamed of cafés on the Bosphorus as snow clogged the streets outside. By spring I was restless, and so when Lloyd proposed a quick three-city tour of Uzbekistan—Tashkent, Bukhara, and Samarkand—I thought, Why not? Lloyd had chosen those cities on the advice of a friend, commissioned the same guide, and not inquired too deeply.

I knew that Samarkand had the glittering Registan, the famous square flanked by tiled facades that look like Persian carpets hung from the sky. Bukhara had the soaring Kalon minaret, so regal that Genghis Khan spared it when he flattened the rest of the city. Tashkent sprawls like Calgary, its broad grid roads, flat buildings, and grit punctuated by massive Soviet hotels. Tashkent is the place you fly into to leave for more interesting places. Tashkent had nothing. Except, I would learn, Tashkent has the blood-soaked Qur'an of Uthman.

If our geography was hazy, we were not alone. It's always been hard to get a handle on Central Asia. In *The Ends of the Earth: A Journey to the Frontiers of Anarchy*, Robert Kaplan quotes an anthropologist who calls the area a "vast shatter zone … where national identities are multiple and polycentric." Yet, says Kaplan, "it is *one* world." The Elizabethans called it Tartary, a place I knew only from literature. (Macbeth's witches use "nose of Turk and Tartar's lips" in their potion. In Dickens's *Great Expectations* the beautiful Estella is described as a Tartar because she's "hard and haughty and capricious to the last degree." Then there was Christopher Marlowe's blood-thirsty play *Tamburlaine the Great*, in which he humiliates and kills the kings of Asia and burns the Qur'an at the end.) Genghis Khan conquered the area, and those Mongols who stayed interbred with the locals. It was the Russians who lumped them all together under the name "Tatar."

The label *Turkestan* is being used again to refer to the whole area, but the designation is still too narrow, because within Turkestan there are not only Turkic peoples such as Turkomans, Uzbeks, Kyrgyz, Kasakhs, and Uighurs, but large pockets of Persian Tajiks and Caucasian tribes, as well as other races. There were few natural boundaries to organize states. What had defined the region was a string of city states like Bukhara and Samarkand, composite entities along the Silk Road, mixing Persian language, Turkic race, and Muslim religion.

We had expected a place fierce and gritty and unrelenting, with a joyless fundamentalist Islam streak underneath. At the airport our guide, Timur, appeared out of the crowd of taxi drivers in an orange T-shirt, jeans, trainers, and a denim sports jacket, with a black nylon bag slung over one shoulder. He had a fleshy face with fashionable stubble and a grin with the hint of a leer. We asked him if he'd be stopping to pray. "Oh no," he said, "I am Muslim, but I am moderate!" At lunch he said, "Mr. Edvard! You must have vodka!" We had a sip; he had a carafe. At dinner he ordered Uzbek *pilau* for us (rice, mutton, vegetables), and for himself a pork chop bigger than a Texas steak. He flung a chunk onto my plate. "Is this not the best pork chop you have ever eaten?"

What surprised us as much as Timur were the pretty girls. Lloyd's wife, Nicole, said, "They all have such slender legs! So beautiful." True, and we could see them. Many wore black skirts cut just at the knee, stylish shoes with block heels, and tight white blouses done up to the neck, but so tight "I'm surprised they can breathe," said Nicole.

A young mother blew Lloyd a kiss. "Is that usual?" Lloyd asked Timur, thinking he'd been extraordinarily favoured.

"Oh yes," said Timur, "Uzbek girls are very friendly." Clearly our view had been skewed by the relentless parade of burkha-clad women in the media.

Timur took us first to Amir Timur Square, where the massive Hotel Uzbekistan ("Typical Soviet architecture") glowers over the statue of Tamerlane. "We have this magnificent statue of Emir Timur, who you call Tamerlane or Tamburlaine. Why? Because after the breakup of the Soviet Union we could no longer have a statue of Stalin or Lenin, so they looked around for someone to be the national hero and they decided on Emir Timur." Rumour had it that a statue of Lenin had been melted down to make this one of Tamerlane. "I myself am named after Timur. Since independence, half the boys in Uzbekistan are named Timur."

Tashkent itself shocked us because it was so clean—cleaner than Geneva. No papers, no plastic bottles, no dirt, though there was construction everywhere. Even the cars seemed clean. We saw one or two ancient Ladas, but unlike in India, where the roadways look like a vehicular night of the undead, there were no wired-together wrecks, no black-belching diesels. Timur said, "Yes! All the city vehicles use natural gas. We have much gas. Eighty-two percent of the known gas reserves remaining in the world right here in Uzbekistan. This is why the Americans love us. Also because we share sixteen miles of border with Afghanistan."

This helped locate me. We were perched above Afghanistan. Timur went on, "There used to be Turkestan—but it was too big, too rich! It would be too powerful! So. Stalin, what does he do? He takes a map, and his pen, and his bottle of vodka, and he drinks and he scribbles lines on the map. When he is done we have Turkmenistan, Uzbekistan, Kazakhstan, Kyrgyzstan, Tajikistan. Everything mixed up." British travel writer Colin Thubron observed that Uzbekistan is shaped like a dog barking at China. *Stan* simply means "place"; Uzbekistan is the place of the Uzbeks—except that it's not. Stalin's ink sliced through ethnic enclaves and divided areas like the fertile Ferghana Valley, which did have a kind of natural unity, between three countries, thus ensuring eternal strife. Like China re-inking the boundaries of Tibet, these arbitrary marks on paper had profound effects on the ground.

I knew we were at the centre of the Silk Road, but even that is a misnomer. Coined in the nineteenth century, the term *Silk Road* actually describes a network of routes that for centuries traversed

Central Asia, from China to Rome. Caravans of up to a thousand camels moved on average twenty-five kilometres a day, often at night because of the heat, prey to brigands, wild animals, avalanches, and sandstorms. They brought silk from China, returned with wine, pomegranates, and innumerable other commodities; they also traded in religions, science, and the secrets of gunpowder, paper, and ink.

Timur flagged down a taxi and brought us to the old quarter; from boulevards as wide as airport runways we turned into narrow streets that wound to a quiet plaza. Turquoise domes glowed against a soft grey sky. The place was deserted except for a couple of kids kicking a ball in the far corner of the square.

"Here is the little mosque, not so impressive," said Timur. "But this is the place of the head of Islam in Uzbekistan. The imam holds his office at the pleasure of the KGB. Our dictator Karimov first thought he could unite the country by promoting Islam, but he soon found out that would cause much trouble. So he has put very strict controls on it. If the imam doesn't go along with the KGB, whoosh! He's out. Still, the library has some nice old books. Come Mr. Edvard, you will like this." We took off our shoes, stepped inside the low building, and found ourselves staring at the Samarkand Codex: the oldest Qur'an in the world. Even for an infidel this was a thrill.

In 651, nineteen years after the death of the Prophet Muhammad, Uthman, the third caliph, decreed that all the fragments of the Qur'an should be collected. Various sections had already been circulating, dictated by the Prophet, who could neither read nor write, or by others who had memorized the texts. Inevitably, discrepancies occurred, and Uthman's aim was to establish a standard edition of the Qur'an. The project took five years. When it was finished Uthman had five copies made and sent them to the major centres of Islam—Medina, Mecca, Basra, Kufa, and Damascus—keeping the original. The metre-high leather pages (deer skin, said one account; gazelle, said another) were yellow and glossy on one side, white and wrinkled on the back. The letters, spare and beautiful in a stately Kufic script, glided across the page in pale brown-black ink. But the context was anything but calm.

Uthman burned the other versions; seeking to impose unity, he only provoked outrage. Supporters of his rival, Ali, burst into his room and stabbed him to death as he was reading his Qur'an. His blood spurted onto the page he was reading, at The Cow 2:137—"And if they believe even as ye believe, then are they rightly guided. / But if they turn away, then are they in schism." Prophetic verses. Any chance Islam might have had for political unity was over. Ali was elected caliph, but he in turn was killed by Uthman's supporters. Thus began the deep Shia–Sunni split that still slices through Islam, a fractured legacy from this book in front of us.

II

Although we use the term *Qur'an* for both the book and the text, *Qur'an* refers to everything that was revealed to and recited by the Prophet Muhammad, and this does not have a physical form. The material object is the *mus'haf* and, like Stalin's ink on the map of Turkestan, the ink of Uthman's calligraphers had fixed the conditions for turbulence. Yet how else to preserve the Word?

The earliest Qur'ans were copied on parchment, which was made of goat, calf, donkey, or even gazelle skin, but the most common was sheep. The calligrapher penned the text freehand on individual sheets, which were then gathered in quires and bound in leather. Most of the early ones were produced in horizontal ("landscape") format—which differentiated them from other non-Qur'anic texts. Scribes would begin with the tenth section (each of the thirty sections consists of twenty pages) and continue to the end; then they would return to the beginning, to ensure that the first sections were written in a perfect script, since any stylistic flaws due to initial awkwardness would have disappeared by then.

The calligraphers used a brown, tannin-based ink: iron-gall ink. Often called the most important ink in history, it is certainly the medium *of* history—Leonardo da Vinci drew with it, Bach composed in it, and Shakespeare wrote with it and punned on it (in *Twelfth Night* Sir Toby urges Sir Andrew, "Let there be gall enough in thy ink …"). The writers of the Dead Sea Scrolls used it, and the American government required it for all official documents up until after the Second World War.

Gallnuts aren't nuts, they're pathological growths, "vegetable antibodies" produced by oak trees in response to eggs laid by gall wasps in the soft tissue of the tree. The "curious excrescences," as Mitchell calls them, develop into homes for the larvae. After five or six months the insects bore their way out. Traditionally, gallnuts that still contained the insects produced the most tannin, and therefore had the highest commercial value. Unscrupulous sellers would sometimes cover up the little holes left by the insect to make it look as if the gallnut still enclosed a larva. The best gallnuts were produced on the branches of small oak, *Quercus infectoria* ("abundant on the Syrian coast, and on the east of the River Jordan," says Mitchell), and exported to Europe, where they were known as Aleppo, Turkey, or Levant galls. They had galls in England, too, but these "oak apples" were literally a pale version of the spiky Mediterranean galls with their high tannin content: Aleppo galls yielded as much as eighty percent tannin, while the smooth Cheshire galls could manage just over twenty-five percent and the Surrey galls in winter sank to eleven percent.

For permanence and long-term stability there was nothing better. The ink's virtue, and sometimes its vice, is that it didn't sit on the surface; it reacted with the collagen in parchment or the cellulose in paper, eating into it, forming a chemical bond. The dark side of this power to bond is that it can destroy what it seeks to preserve, leaving brown-haloed holes where once were musical notes, poetry, or deal-sealing signatures. This was because a byproduct of the reactions between the iron sulphate and the tannic acid of the gallnuts was sulphuric acid, which would become more concentrated the longer the ink was exposed to air. It feels magical to write with gallnut ink because it goes on almost clear, like water, and then a few inches behind your pen the fluid darkens as it catches up with you, deepening from the merest beige to a warm brown, the colour of a perfectly toasted campfire marshmallow. After a few hours, or a few days, the ink may turn black. Or, as with the Uthman Qur'an, it may stay brown around the edges.

In 1945, at the same time that Bíró's ballpoint was flying high with bomber crews, the United States imported more than 550,000 pounds of Aleppo galls from Turkey. This most ancient of inks was still being used for official documents. In David Carvalho's wonderfully obsessive, cranky *Forty Centuries of Ink* (1904), he observes that gall ink, like paper, entered the West at the end of the eleventh century, making its way from Asia through Arabia, Spain, and France. It came into vogue in the twelfth century, when it was adopted by the Catholic Church; from the priests in Rome it flowed out across Europe. But Carvalho emphasizes that while the *Secreta* of Theopholis (which contains the "gall" ink formula) is Italian, "the invention of this ink belongs solely to an Asiatic country." Central to Carvalho's book is a diatribe against additives.

The issue is complicated, but in brief, inks that worked fine with quill pens did not work well at all with steel nibs, which would clog with the sediment that naturally formed in gall inks. The acid that was added to get rid of the sediment also ate away at the paper.

In 1891 an investigation of official U.S. birth, marriage, and death records discovered that some of the old inks had "eaten through the paper as thoroughly as if the writing had been done with a sharp instrument." Carvalho cites court cases where citizens had lost property because wills and real estate transactions had become illegible. The problem was that ink with too much acid would burn through and ink without it could be washed away. Carvalho rails against the NYPD blue: "In the police department of this city a blue is often used which is made from Prussian blue…. A little soap and water will wipe out this writing, so that the record can be easily altered at any time. The use of this ink in the police department is said to date from the time of Tweed…." William Marcy—"Boss"—Tweed, head of Tammany Hall, the Democratic party political machine, stole over a hundred million dollars from New York City; he died in jail, convicted of larceny and fraud. The bad ink was an emblem of his corruption. In 1894 Carvalho introduced a bill that would provide for an official ink to be used for all New York state documents, an ink that would secure uniformity and "throw many obstacles in the way of altering the records." The bill failed.

Today the issue of data corruption is even more urgent, more pervasive, and not just a question of wills and real estate deeds. In warm, humid conditions, CDs and DVDs experience literal rot, and if they haven't been burned at the slowest speed (and stored in the dark, and at the proper humidity), they'll rot faster. This is apart from physical abrasions like scratching. If you're lucky, they'll last fifteen years. Some technicians say six. Then there is "bit rot":

digital files degrade as they lose their magnetic orientation, and the magnetic bit strength of hard drives gradually fades as well. Technicians advise recopying to a new format every five years. They point out that the computer industry is dedicated to planned obsolescence; one writer recommended that if you really want to preserve a document, use paper and ink.

How many times would the Qur'an have had to be transferred if Uthman had compiled it into a digital file? Timur wasn't interested in old books, and chatted with the unarmed security guard, who provided little security. What I thought would be an object of international reverence was for them a curiosity in a Plexiglas case. Timur said, "Our dictator Karimov does not promote this. It could become a place of pilgrimage, but he does not want the Islamists here." If the Samarkand Codex were a flash drive under glass, that wouldn't be a problem.

III

We took the SHARQ, the fast train from Tashkent to Bukhara, zooming across the salty plain. It was like crossing Saskatchewan in springtime—white patches on a flat brown landscape, except that the white patches weren't snow but salt deposits. In Bukhara we stood at the base of the Kalon minaret and imagined what it would have been like to have been a convicted thief thrown in a burlap sack from the top of the forty-eight-metre-high "Tower of Death." Genghis Khan, perhaps seeing its value for dramatic public executions, had spared it, one of the few things he did spare in 1220 CE when he sacked the city. In my brother's French guidebook I found this wonderful sentence: *Sous le soleil, la plaine miroitait, du sang versé.* The English version, "From the reflection of the sun the plain seemed to be a tray filled with blood," lacks the triple rhyme of the French, and that fine verb *miroiter* (to shimmer, sparkle), but it still captures the scene: thirty thousand Bukharan troops slaughtered by Genghis Khan outside the gates.

We drank green tea and ate rounds of crisp *non*, the unleavened bread, beside the Lyab-i-Hauz, the pool surrounded by cafés that forms the centre of the old town. In this placid setting I read of Bukhara's tortuous politics during "the Great Game," the period of rivalry between Russia and Britain for influence in Central Asia— of Nasrullah "The Butcher" (1874–1920), who murdered his three brothers and on his deathbed ordered the murder of his wife and daughters to ensure their chastity, and of the British spy, Fred Bailey, who in 1919 was engaged by the Bolsheviks to track himself down, on which mission he entered Bukhara carrying false identities and a developer for invisible ink. Later we visited the Emir's summer palace, where peacocks screeched and flashed green feathers the colour of the ink in my copy of the *Rubaiyat*, and where a young artist sold miniatures of the legend of Husrev sighting Shirin naked in a pool, done in the style of Bihzad, the famous fifteenth-century Persian painter.

I had encountered the scene in Orhan Pamuk's *My Name Is Red*, where the dispute over adhering to Bihzad's style or allowing Western influences leads to murder. Then, on a trip to New York, I convinced the curators at the Met to let me view Bihzad's painting. It's very stylized, but still very sexy. The sky behind Husrev is gold, the water black (painted in real silver, it would have soon tarnished). Shirin's red top is lying beside the pool, and she's still wearing her blue pants. Her long black braids come down over her breasts. Husrev sits on a mottled grey horse, his finger in his mouth in the traditional gesture of surprise. This is the first glimpse the lovers have of each other; the murders and suicides are yet to come.

An ancient story made famous by the Persian poet Nizami in the twelfth century, the tragic romance is as well known in the East as Romeo and Juliet is in the West. It continues to be retold and adapted, most recently in a Los Angeles–based romantic comedy *Shirin in Love*. Proto-punk musician Jonathan Richman references the story in his 1985 song "Shirin and Fahrad," but outside of Pamuk's novel, I'd never heard it.

For my generation, the only contact with Islamic culture, aside from my grandmother's "Turkish" carpet, was likely to be through such works as the *Arabian Nights*, filtered through British translators, or such poems as "The Sick King in Bokhara," by Matthew Arnold, who got most of his details from *Travels into Bokhara* by Alexander "Bukhara" Burnes, diplomat/spy cousin of Robbie Burns. Arnold had never been there, and neither had Edgar Allan Poe, but that didn't stop him from writing the bombastic "Tamerlane" (he should have stuck to ravens). James Elroy Flecker, in *Hassan: The Story of Hassan of Bagdad and How He Came to Make the Golden Journey to Samarkand*, at least made the city sound salacious:

> We travel not for trafficking alone;
> By hotter winds our fiery hearts are fanned:
> For lust of knowing what should not be known,
> We take the Golden Road to Samarkand.

Unfortunately his characters never get there. I slumped in the back of the minivan, half-snoozed under the Uzbek pop music, and watched the arid landscape pass. The Golden Road was much more exotic in the work of the poets.

I remembered my father quoting from the Bible something about the moving hand that having writ moves on and how nothing can erase those words. It made us kids shiver when he chanted it and waggled a finger at us—he had very bony fingers. It was years before I discovered that the lines weren't from the Bible but from Omar Khayyam's *Rubaiyat*: "The Moving Finger writes; and having writ, / Moves on, nor all thy Piety nor Wit / Shall lure it back to cancel half a line, / Nor all thy Tears wash out a Word of it."

My father had his father's copy, and when I was ten or eleven I was allowed to take it down and look at it by myself. None of the other books in my father's library felt like this. It's a small book, oddly square, with brown suede covers that droop over the pages. Centred on the front is a smooth-leather panel with embossed images of a crescent moon, six stars, a wine jug, and a goblet. I loved the feel of the soft leather cover and, on the thick pages inside, the indentations for the initial capitals, alternating in red or blue ink. On the first page in green ink is a peacock in front of domed buildings; at the back, also in green, is an upended empty wine glass. I see it now for what it was—a slightly pretentious middlebrow production—but for a boy of ten or eleven it was an exotic object. I never dreamed I'd see those green-inked domes.

My grandfather had scrawled his name, E.T. Bishop, at a forty-five-degree angle in the upper right corner of the loose endpaper. He wrote in black ink with a fine nib, no doubt with a dip pen of the kind Marcel Bich would start out making. The capital *B* flows like an open heart off the crossbar of the capital *T*. The lowercase *s* and *p* aren't closed—the handwriting of a casual young man. But he's also earnest (he's a new lawyer from Osgoode Hall, after all), and he has a new stamp to emboss his Ex Libris, and so he stamps the title page, the back page, and one page in the middle—thus no doubt deterring thieves but marring the book in the process. Otherwise he's careful, marking certain verses with small pencil ticks, such as this one in number VIII:

> The Bird of Time has but a little way
> To Flutter—and the Bird is on the Wing …

and the next one:

> The Wine of Life keeps oozing drop by drop,
> The Leaves of Life keep falling one by one …

Verses appropriate to that period of life when time seems most fleeting—youth gone, ambition empty—your late twenties. But he was austere, this young man, no tick marks on those verses about embracing wine, though he did mark the famous quatrain:

> A book of verses underneath the bough
> A loaf of bread, a jug of wine and Thou—
> Beside me singing in the wilderness
> Oh, wilderness were paradise enow!

Later I would learn about the importance of diacritical marks in Arabic: the word for "book"—*kitab*—with a change of one dot becomes *kabab*, as in *shish kabab*—"roast meat." One killjoy scholar argues that *kabab* makes more sense in the context, turning poetic romance into mere lust on the grass.

The book had a colophon at the back, full caps in red, that announced that this *Rubaiyat* had been "done into a printed book by the Roycrofters, at their shop in East Aurora, which is in Erie County, New York, in April MCMVI." The Roycrofters were a group assembled by Elbert Hubbard, a successful soap salesman turned writer who, frustrated because he couldn't find a publisher for his biographical sketches and inspired by William Morris, set up his own press and named it after Samuel and Thomas Roycroft, who'd been printers in London at the same time as Moxon. My younger self thought "done into a book" sounded old and mysterious, not pompous; my older self liked the connections. East Aurora was forty miles south of where my Uncle Jeremiah had trained. They picked the right time to do their *Rubaiyat*. When Edward Fitzgerald made his translation in 1859 it sank like a stone. Issued in an edition of 250 copies, even that didn't sell out. Then it was translated into French, and the French loved it—it was daring and decadent, calling for wine and flirting with atheism—and now, with the stamp of French approval on it, the English and the Americans too embraced the *Rubaiyat*. "Omar" societies were formed, and in the 1890s hundreds of American children were called "Omar."

Timur, who talked politics constantly, neglected to mention that the main cities of Uzbekistan, Samarkand and Bukhara, are predominantly Tajik, not Uzbek. *The Rubaiyat* too is Tajik. Omar Khayyam lived in Samarkand (Timur didn't tell us that either); he was a scientist who developed mathematical theories that were way ahead of their time and who experimented in weather forecasting. But, as Amin Maalouf makes him say in his novel *Samarkand*, "The theories they construct themselves will inevitably be destroyed and even ridiculed by those who come after. That is the law of science. Poetry does not have a similar law. It never negates what has come before it and is never negated by what follows. Poetry lives in complete calm through the centuries. That is why I wrote my *Rubaiyat*." Rubai (Persian quatrains) weren't high-class like haiku, they were low like limericks, and in *Samarkand* Khayyam's beautiful poet-lover Jahan expresses concern that he's wasting his time and his talent. We readers, from our superior historical perspective, know she's wrong.

In its nineteenth-century translations, *The Rubaiyat* flirted daringly with hedonism and agnosticism, extolling the joys of wine, celebrating the pleasures of the moment, and questioning the spiritual life. My grandfather had ticked LXIII:

> O Threats of Hell and Hopes of Paradise!
> One thing at least is certain—This Life flies;
> One thing is certain and the rest is Lies;
> The Flower that once has blown for ever dies.

Yet Khayyam's rubai were written in no particular order. Fitzgerald even fitted parts of different verses together, so much so that some commentators refer to his translation as "FitzOmar." Though many of his verses seem to contradict the teachings of Islam, in 1903 Robert Arnot brought out a collection entitled *The Sufistic Quatrains of Omar Khayyam*, which included translations by Edward Fitzgerald, and others. Critics today are still divided, with some claiming Khayyam was an agnostic and others insisting his work is firmly in the tradition of Sufi philosophy.

IV

That night the BBC World Service reported an uprising next door in Kyrgyzstan: one hundred killed, the president's house in flames, rebels in control of the state television station. Should we be concerned? Timur took us to a fancy restaurant and ordered heaps of chicken that we ripped apart with our fingers. He poured himself more vodka and said, "Don't worry! Nothing will happen here! In fact there are a lot of Uzbeks in the south, and they are the wealthiest group. In fact they probably support the uprising."

"So you mean it's not a spontaneous uprising over the price of bread like the BBC said?"

"Ah, Mr. Edvard! Spontaneous uprising! It is never a spontaneous uprising! That is a fantasy, like Santa Claus." Two days later, 500,000 refugees fled and the Kyrgyz troops sent to keep the peace attacked Uzbeks in their own neighbourhoods. The joke about the vodka-drawn map lines didn't seem so funny.

The next day I forgot about Kyrgyzstan—I had my own cultural uprising. Timur took us to the central market: flowers, sweets, nuts, and mounds of spices … he stopped by a heap of green powder. "Here is *nos*. We like it very much here in Uzbekistan. Come, Mr. Edvard!"

Timur scooped some up on a spoon, then knocked half of it off. Not a big dose. He showed me how to put it in my hand, and then changed his mind and told me to open my mouth. I curled my tongue back and he poured it in. I clamped my tongue down. At first I felt nothing, then it began to feel hot, like chilies, and saliva filled my mouth. "Can I spit?" "No, hold it for a few minutes." We walked toward the van and my mouth felt like an overflowing bathtub, leaking at the edges. I leaned against a minaret and spat. Now I was dizzy.

"Ouf. This stuff is powerful," I said. I felt stoned. My brother told me later that I'd turned an interesting grey colour, but he hadn't wanted to alarm me. I staggered as I got in the van.

"How do you feel, Mr. Edvard?"

"Okay," I lied.

"It is Uzbek cigarettes," said Timur. "It will pass quickly."

When I smoked my first cigarette as a kid I felt dizzy and then slightly sick. This felt as if I'd smoked a whole pack of cigarettes, maybe a whole carton. The motion of the van was making me sicker. My vision contracted, the tunnel vision of nausea, and I slumped in my seat. I could feel the cool air from the window on my face, but I was sweating. I watched the droplets of sweat fall off my forehead onto the floor. I couldn't look out the window now, couldn't raise my head, and wondered how to avoid throwing up on the side of the van.

"You need to eat, Mr. Edvard!"

I didn't think so.

We stopped at a lunch place.

I lurched out onto the road, gorge rising, tea-rice-bread filling my mouth, spewing out through my hand trying to hold it in, then into the ditch in front of the restaurant. The sun felt good on the back of my neck as I retched again, leaning on the retaining wall. Lloyd brought me a rag to wipe my face and poured water on my hands.

I went back to the hotel while the others toured the glorious tomb of Timur. The ride was a blur and I had to crawl to the bed. Part of my brain watched me crawling across the floor, and said, *Well, you wanted an authentic cultural experience.* The disorientation of Central Asia was no longer abstract. I fell into a deep sleep, but after forty-five minutes my eyes snapped open. I swung out of bed, ready, with that peculiar clarity you have after illness, to see the Registan, and to learn how Uleg Beg lost his head.

Before coming to Uzbekistan I hadn't heard of the man, which is like never having heard of Galileo. Timur did tell us about him: Uleg Beg was Tamerlane's grandson, and grew up in his court. In 1424 he began work on an observatory. He assembled some sixty scholars to make instruments and work on mathematical formulae; twenty years later he issued a set of astronomical tables that plotted the coordinates of 1018 stars and measured the stellar year to within a minute of modern calculations. His work was used throughout the Islamic world and translated into Latin. Uleg Beg was born in 1395, the same year as Gutenberg, which gave me a necessary point of reference—not only did Samarkand seem unreal and Central Asia vague, but everything out there seemed to take place in the indeterminate time of the tales in the *Arabian Nights*.

Timur brought us first to the courtyard of the Bibi Khanum mosque. He showed us the Bibi Khanum minaret, where Tamerlane had thrown his wife off for allowing the architect to kiss her (the architect, legend has it, grew wings and flew to Mecca). Timur then sat us down in the shade beside a massive marble lectern that once held the Qur'an of Uthman. The Caliph Ali had taken it to Kufa, and when Tamerlane invaded Iraq he brought the Qur'an that we'd seen in Tashkent back to Samarkand. Tamerlane was bloodthirsty but valued the arts; he also brought Egyptian and Syrian book-binders to Samarkand, initiating what is seen now as a golden age of book production. Uleg Beg donated the lectern, and displayed the big Qur'an to the public on special days and to pilgrims who came to worship the book—the practice that Uzbekistan's dictator, Islam Karimov, feared. It stayed in Samarkand for four hundred years, becoming war booty again in 1868 when the Russians took it to St. Petersburg. After the 1917 Revolution, in a gesture to consolidate relations with the Islamic provinces, Lenin returned the Qur'an and it eventually found its way to Tashkent.

We crossed to the Registan, what George Curzon called, even in ruin, "the noblest public square in the world." Here Uleg Beg created a caravanserai, a hospice for dervishes, and the world's best university. The trees in the inner courtyard had new blooms, a fresh breeze was blowing, birds were twittering, tourists murmuring. A little boy with a chunk of bread stomped along the big paving stones, stretching to avoid the cracks. Uzbek girls in bright red skirts walked by. There was an inscription over the gate—*It is the sacred duty of every Muslim man and woman to pursue knowledge*—but of course, in Uleg Beg's university, only boys would have been allowed, and few enough of them: just two hundred students and the most accomplished scholars in the world.

I thought of the students going to the little classes on the ground floor, where now merchants urged us to come in and look at their rugs and scarves and trick pencil boxes. Each door was set into a high arched alcove. On the courtyard's walls blue and turquoise tiles were interlaced with lines of gold against a tawny background the colour of damp sand. Even in the heat it would feel as if you were surrounded by water. A truly fabulous place to study, knowing that you were at the very centre of learning and beauty. Did they realize their secular studies were generating hatred?

Although Uleg Beg donated the lectern for Uthman's Qur'an, he declared, "Religions dissipate like fog, kingdoms vanish, but the works of scientists remain for eternity." Maybe so, but scientists themselves are vulnerable to religion. (Khayyam bet on poetry, which proved safer.) Uleg Beg's elder son, who resented his father's interest in secular learning (and felt that Dad favoured the younger brother), had him tried by a court of dervishes, who sentenced him to make a pilgrimage to Mecca, and then had him beheaded as soon as he left the city. They burned his observatory to the ground; all the scholars fled. It's fruitless to indulge in the "What if?" school of history, but I couldn't help thinking, What if this madrassah had, like the universities of Oxford or Bologna, continued from its inception to the present? It was 1449. As the son's henchmen were hacking off his father's head, Johannes Gutenberg was perfecting his thick ink in Mainz. The centres of intellectual inquiry were beginning to shift.

ULEG BEG WOULD HAVE fit better in the contemporary regime, which encouraged secular learning. For our last excursion Timur took us up to the village of Tersak, in the hills above the plain, to meet a retired French teacher. He told us that the province is divided into various divisions; in each you learn Russian first as a foreign language and then one of French, German, or English, depending upon the division. We toured the school, which was both strange and eerily familiar. The students sang "Frère Jacques" and "Au Claire de la Lune," and on the wall, along with pictures of Karimov and drawings of Emir Timur, they had a chart for cursive handwriting like the one in my elementary school. I'd hoped to see inkwells, but it looked as if the school supplied cheap ballpoints.

The teacher wore a dark-blue caftan over a western-style shirt and pants; his grey-black hair was short and freshly cut, a tan line visible around the back of his head. He looked fit. When he was a student he used to bicycle twelve kilometres each way, and he still liked to ramble around the hills. He loved Maupassant and Zola. He had eleven children and many grandchildren about the place. Under the Soviets a woman got a gold medal if she had more than ten children; his wife had gotten one, which he melted down and used to coat his teeth. He flashed a golden smile. He'd studied in Paris but loved Tersak. His daughter took us on a hike to a waterfall above the village, banging a stick to scare the snakes off the path. Although the way was hot and dusty, little children came to the edge of their yards to call *"Bonjour!"*

When we got back, the schoolteacher said, *"Vous avez fatiguez, monsieur?"*

"Non," I said, unwilling to admit it, but conceded, *"Je suis très chaud."* Even as I said it I knew I was making an error—Lloyd told me later that I'd said I was really sexy—but the teacher simply said "Ah," and poured the tea.

When the Uzbeks make tea they first pour it into a cup and back into the pot three times. The French teacher did this with great flair, starting with the spout near the lip and then raising the pot up as if he were pulling the column of tea out of the cup, and then quickly lowering it just before it overflowed. Then he put his left hand over his heart and passed the tea with his right. We were sitting around a low table on wide benches with our legs folded, eating pilau, and the sun was very hot. Occasionally a breeze would come down the valley. You could see snow patches bright against the green of the upper slopes; the breeze carried their coolness. In the summer it's forty-five down in Samarkand, said our host, but twenty-five up here. In a few weeks the vines on the pergola over the table would burst into leaf and there would be shade.

He asked me what I did. Brave now with my undergraduate French, I told him, *"Je suis un professeur de litterature. Et je fais des recherches pour un histoire de l'encre."*

"De l'encre?"

"Oui."

"Ah," he said with a smile of gold-capped teeth, and told me of how as a boy he carried a schoolbag with an inkwell and a box of pens. He started out with a wooden pen holder into which you fitted a nib.

"Moi aussi," I said. Here we were, two old men of the pen.

V

Apart from its Registan, the only thing I'd known about Samarkand before I left home was that it was important for paper: in the battle of Talas River between Chinese and Arab armies in the summer of 751 CE, the Arabs won a decisive victory over the Tang army. My favourite Chinese poet, Du Fu, was on the other side of this, writing his "Song of the War Carts," lines that made me think of the current American president: "and then they ship us out again / to where the blood is lapping like the sea / the emperor wants to expand his realm / … boys seem born to die in foreign weeds / … the empire isn't big enough?"

The aftermath was more significant than the battle. The Arabs took prisoners who bartered their knowledge of paper making, until then unknown in the West. Samarkand became a centre of paper manufacture, and for hundreds of years the best paper was known as "Samarkand paper." The technology of papermaking revolutionized the Islamic world, and later Europe—much later: by 794 there was a paper mill in Baghdad, but it wasn't until 1151 that the Moors set up the first paper mill in Europe, in Xativa, Spain. The Christians persisted in thinking that papermaking was witch-craft. We got our inks, our paper, and our learning from Central Asia, and I had only the vaguest sense of it. Even our "loaf of bread, jug of wine, and thou," and my father's moving finger, came from here too, but by way of a British translator, nothing ever said of the original poet. It was as if we couldn't conceive of there being anything out here.

During the winter I'd read of a ceremonial procession in Egypt in which "the canopy of the caliph was preceded by a sword and a magnificent inkwell as the symbol of the 'People of the Pen'"—the military caste was known as "People of the Sword" and the class of scribes as "People of the Pen." There was a special gallnut ink for the People of the Sword that uses green vitriol and yellow myro-balan (an astringent, plum-like fruit). My book didn't say what the ink was for the People of the Pen. Maybe all the inks were for people of the pen. I wished my French were good enough to ask my host in Tersak about inks, for where I had Waterman's Blue-Black, and Parker Quink, with turquoise the only variant (and only used by girls), I knew that Islam had a rich heritage.

Around 1025 Al-muizz ibn Badis, an eighteen-year-old prince in what is present-day Tunisia, produced *The Staff of the Scribes*: a treatise on inks and writing implements. It opens with a quotation from The Clot 96:3–5 of the Qur'an: "The exalted One said, 'Read, by your generous Lord who taught by pen.'" Ibn Badis continues, "The first significant thing that Allah created was the pen. And he said, 'Flow,' and it flows with whatever it is, until the day of resurrection." Ibn Badis links calligraphy with clarity and truth: "The prophet, may Allah bless him and give him his peace, said, 'Beautiful writing gives to truth more clarity. It demonstrates that when the pens are good, the books smile.'" For the pen is more than a passive instrument: "The pen is the molder of speech. It molds whatever the mind contains. Anything which the pens have given fruit the ages have not dared to erase. But the pen is a tree and its fruits are the words and the thought is the pearl of wisdom."

If the thought is not to be erased you have to have good ink. One ninth-century poet exults in "ink like the soaring wings of the raven, papyrus like the lustrous mirage." Ibn Badis mainly used the gallnuts of terebinth (a shrub that was the source of turpentine) and tamarisk, a bush with feathery branches and tiny pink flowers that grows in sandy soil in Western Asia, Europe, and even England, where, according to Mitchell, the terebinth galls were called "the galls of Sodom"—but still coveted. (In 1609 an iron-gall ink was invented by Guyton and sold on the Pont Neuf in Paris under the title *Encre de la petite virtue*, one more perpetuation of Western views of Eastern sexuality.)

We're lucky to have the recipes of ibn Badis, because many calligraphers had their own secret formulae that died with them, and ibn Badis has wonderful inks: an ink that can be used immediately; a dry ink for travellers; a cheap ink for the common people;

an ink that doesn't need fire for its preparation; an ink for religious books; an ink in which pomegranate rind is used in addition to gallnuts. For Kufic ink,

> Take the rind of pomegranates and procure wood to burn it.
> Take the ash and knead it with yogurt and a little of the
>   moistened gum.
> Then bake it into cakes and dry it in the shade.
> This is then the best type of ink.

The reference to "shade" reminds us that we're not in a climate-controlled laboratory—elsewhere we are advised to mix our gallnuts and gum arabic with water and leave it in the sun for three days, then leave it in the sun again, for four days if it's summer, twelve days if it's winter. And the instruction to "procure wood" was a reminder that fuel was expensive in many parts of the Middle East. Kufa, the centre for learning—where Kufic script, the first calligraphy style for the Arabic language, originated in the seventh century—is 170 kilometres south of Baghdad, on the Euphrates but surrounded by desert.

I learned to read the recipes of ibn Badis slowly, like poems, and discovered, as in Cennino Cennini, the sense of time within them. Iraqi ink is even more extravagant, combining as it does the floral and the fecal:

> Anemones are taken and stuffed into thin vessels and buried
>   in the dung of asses until melted, watery, and dissolved.
> Then paper sheets are burned. What has been burned is
>   gathered with the liquid and removed to dry in the
>   shade.

> Then a dirham of it is taken, a dirham of gum arabic,
>   and one-half dirham of pulverized gallnut. These are
>   pulverized together with white of the egg. It is made
>   into a ball.
> Then it is dried as has been mentioned, put into the ink-
>   well as needed with water of sorrel. This is the best
>   water for it.

The terms alone are fantastic: King's ink is a special ink made "from soot of refined storax, soot of the sandarac, and soot of laudanum—either together or separate." *Storax*, the *Oxford English Dictionary* told me, was used as an aromatic for thousands of years; originally resin from the styrax tree (which sounded like something from Dr. Seuss), it now referred to "any of several fragrant resinous exudations." *Sandarac* is a coniferous tree that grows not only in North Africa but also in Australia; and *laudanum*, I knew, was the opiate that transported Coleridge to an imaginary Orient for his poem "Kubla Khan." In making Persian ink you are instructed to "lute the vessel with the clay of the art": lute is potter's clay, and to lute is to make airtight with clay.

The coloured inks have fabulous names: "yellow apricot"; "pomegranate blossoms"; "blood of the gazelle"; "colour of dates beginning to ripen." And there are the "colours of the wild"—red earth, white lead, and a little yellow arsenic: "This colour is of the fresh wild animal" (Ibn Badis is talking about the meat, not about the coat). "If the color of the lion is wanted then a little bit of *lazward* is added … if the colour of the falcon is desired, then the required amount of *bauraq* with a little yellow arsenic and arsenic of the black type." There is peacock red, and ruby red, rose-coloured ink, purple ink, pistachio ink … and if you are very rich you can write with gold:

Pure gold is beaten to a thin leaf.
It is then cut up into small pieces. On it is poured borax.
The fire is then introduced and blown on it until melted.
It is then rubbed with a stone until it becomes like butter.
It is then gathered and pressed until the liquid comes out
    and the gold remains …
Write with it like ink. It is good.

If you want secret ink you'll need a wild ass and a camel, but as a bonus this ink will cure your liver ailments:

Take two dirhams of goat yogurt and two dirhams of milk of
    the wild ass; throw this into five dirhams of grape juice
    and leave it for ten days.
Then dissolve it in fifteen dirhams of camel's milk—that
    camel whose whiteness is inclined to redness.
Then write. It cannot be read except in the light of the lamp.
If a man has yellow jaundice and drinks one-half dirham, he
    recovers.
It is also for those who have liver fever.

Is it the camel's *milk* that inclines to redness, or the camel itself? In any case, there are twelve dirhams to the ounce, so you don't need much. Ibn Badis offers a less complicated recipe with cow's milk, but you have to leave it for forty days. A long time—so not the thing for sudden political intrigue or a surge of illicit passion. *The Costly Pearl* by 'Abd Rabbih (860–940) offers an intriguing variant: for ordinary secret messages use milk, but "if you desire that the writing shall not be read during the day but shall be read at night, then write it with gall of the turtle."

These ink recipes from a thousand years ago have the density of poetry; they take you into a strange world where men and women grind the rind of pomegranates and conduct love affairs with ink from the gall of the turtle. As we flew back out of Tashkent I decided my encounter with *nos* had been a kinetic metaphor for the intoxication and disorientation of Central Asia (I had to dignify the experience somehow). I come from a province laid out in grid lines, with natural definition—mountains to the left, prairie to the right—and maybe geography gets into your bones and your brain. Everything had blurred in Uzbekistan, and yet from time to time a flash of the familiar emerged from the foreign—Timur with his pork chops, or the French teacher with whom I shared a writing heritage. The turquoise domes of Samarkand turned out to be the endpoint of a line that started with the green ink in my grand-father's *Rubaiyat*. Things were not as separate as I thought. Back in Canada, as I read further into Islam, into disputes in which blood and ink flowed together, the neat categories of sacred and secular, erotic and spiritual, proved as porous as the boundaries of Tartary. Even "home" and "away" seemed less like binary opposites than the curved commas of the yin and yang symbol.

*"Guardians of Revolution" from Shirin Neshat's "Women of Allah" series,
in which she inked Persian verses by radical women poets onto black
and white photographs that often combine guns and the female body.
Shirin Neshat,* Guardians of Revolution (Women of Allah Series), *1994.
RC print & ink (photo taken by Cynthia Preston). Copyright Shirin Neshat.
Courtesy of Gladstone Gallery, New York and Brussels.*

Women of Allah

I

> I watched the tense quivering of my beloved's angry letters,
> the somersaults they turned trying to deceive me with their
> hip-swinging right-to-left progression.
>
> —ORHAN PAMUK, *MY NAME IS RED*

Sometimes a book is your best portal into a culture. In my early twenties I had travelled uncomprehendingly from Istanbul to Herat, Afghanistan, on third-class trains and no-class buses. I got a sense of the landscape and the people but learned nothing of the arts, though I toured museums and gazed dutifully at some ruined towers north of Kandahar. It was decades later, still wondering what I'd seen and now interested in the culture of the book, that I took up Orhan Pamuk's *My Name Is Red* (translated in 2001). On this sprawling novel Pamuk hangs all manner of digressions—mini-essays on the role of the artist and the influence of the West; excursions into the stories of Shirin and Husrev and other traditional tales; large chunks of Ottoman history. I learned that the dusty backwater I'd visited, Herat, had once been renowned for its arts.

Tamerlane's son, Shah Rukh, was sent down from Samarkand to make the city his capital, and under him Herat became a centre for culture, the book arts in particular. His son Baisunghur, himself a talented calligrapher and painter, founded a library and a workshop with forty artists, calligraphers, and painters. It was out of this atelier that the great Persian painter Bihzad (c. 1450– c. 1535) emerged. Pamuk's encyclopedic novel is also a homage to Bihzad, and the book, like Istanbul itself, provides a bridge from West to East (there were no books about Bihzad in our otherwise comprehensive university library). In any case, I was hooked— Bizhad's illustrations would have come overland on the route I'd followed, an astonishing four thousand kilometres by horse and camel, from Herat to Istanbul, where they were incorporated into volumes for the sultan.

In *My Name Is Red* letters are erotic and ink is bloody. At issue is the incorporation of Italian style into Islamic painting, especially the use of perspective. More than a question of mere craft, perspective is neither politically nor theologically neutral, for it means you're seeing from the point of view of man, not of Allah. From the viewpoint of God one sultan looks like every other, but the new sultan, admiring the portraits from Venice, wants to be individualized. His vanity is a challenge to the tradition of Bihzad and to the structure of society. The calligrapher-painters must be killed.

One of the finest set pieces of the novel is the next murder: this time of the head miniaturist, who dies slowly, skull crushed by a three-hundred-year-old Mongol inkpot. Before the blow he and his murderer discuss the "mysteries of red ink, whose consistency he could feel as he gently swung the inkpot," how Mongols brought the secrets from Chinese masters to Khorasan, Herat, and Bukhara.

After the stroke the murderee, head broken, blood oozing, reflects, "What I thought was my blood was red ink, what I thought was ink on his hands was my flowing blood." His body turns bright red from the ink and blood. The author recalls the siege of Baghdad in 1258 when the Tigris ran red from the ink flowing out of the books thrown into the river by the Mongols as they sacked the city and destroyed the great libraries. A grand civilization bleeding out.

At one point in the book the colour red itself (there are twelve different narrators, several inanimate) describes how the miniaturist "furiously pounded the best variety of dried red beetle from the hottest climes of Hindustan into a fine powder." This red beetle is the "common shield louse," about one-third the size of a ladybug. The females are wingless and spend their whole lives clustered on the host plant while the males fly around and impregnate them. The females are harvested when the male insects are "engaged in their erotic seasonal flights" (as an early writer chastely puts it). The males live only half as long as the females, but it is the females (who look so much like berries on the branch that the name *coccus*— berry or acorn—was attached to them) who are crushed for their pigment. As Rosamond D. Harley writes in *Artists' Pigments c. 1600–1835*, when they're harvested "the animal origin is virtually unrecognizable because it is a solid substance made up of the bodies of females, which are dead, with some 200–500 unhatched eggs, all surrounded by a brown-red hardened exudation."

There are three related reds: kermes, lake, and cochineal. *Kermes* (from the Arabic *kermez*, from which *crimson* derives) is made from the *coccus ilicus*, found on the Kermes oak. But the dyes from this insect aren't particularly brilliant, and what Pamuk's craftsman is pounding is the more desirable, more expensive *coccus lacca*, which grows on fig trees in Asia and India. It was sometimes imported to England on sticks (stick lac, or lake), in cake form, or in thin pieces (shellac). Some sources advise the mixing in of a little ear wax, probably to improve the flow of the colour.

Then there is cochineal. The *cochineal cacti* lived on various varieties of cactus in Mexico and South America. The acid content in cochineal seems to have been higher than in insects from the East, and whereas lac from Asia has a purple cast, the lac from Latin America is a truer red. It was intensive to make: the live females were shaken off shrubs, smothered and killed, then dried in the sun for four or five days, during which they lost a third of their weight. It took seventy thousand insects to make one pound of dye. Aztec doctors mixed it with vinegar and applied it to wounds; cooks used it to colour tamales; prostitutes stained their lips with it; Aztec scribes used it in their books.

When the Spanish conquistadores found Aztecs selling a red dye in Mexico in 1519, they knew they had a valuable commodity. In the reign of Henry VI of England a yard of the cheapest scarlet cloth cost more than a month's wages for a master mason; the best red was twice as much. The Spanish, on pain of death, kept the secret of cochineal for two hundred years: no one knew it was an insect. They controlled the supply, and it was wildly expensive. British privateers attacked Spanish ships on their way home from their Latin American colonies (Johnny Depp in *Pirates of the Caribbean* would have been after cochineal as well as gold). Robert Devereux, the Earl of Essex and Queen Elizabeth's young lover, captured a ship with twenty-seven tons of cochineal—the largest cochineal prize of the sixteenth century: it was worth £80,000 at a time when a clergyman made £80 a year. Poet John Donne had shipped with Devereux and writes with mordant wit about cochineal ("Cutchannel") and sex in Satire IV:

Now,
The Ladies come; As Pirats, which doe know
That there came weak ships fraught with Cutchannel,
The men board them; and praise, as they thinke, well,
Their beauties; they the mens wits; Both are bought.

(*Mordant*, by the way, comes from the Latin *mordere*, meaning "to bite"; it was a binding agent in ink or dye, so called because it bit into the fibres.) Cochineal was used in cosmetics as well as ink, and it was even considered an anti-depressant: John Gerard's *Herball*, a seventeenth-century medical text, said it was "good against melancholy diseases, vaine imaginations, sighings, griefe, and sorrow with manifest cause, for that it purgeth away melancholy humors."

When chemical dyes were invented in the early nineteenth century the price dropped and cochineal became available to commoners. In Edith Nesmith's *The Railway Children* (1906), the characters talk about making icing for a birthday cake using egg whites with a few drops of cochineal; and in Noel Coward's *Private Lives* (1930), Elyot tells Sibyl, who has lost her lipstick, to send down to the kitchen for some cochineal. In the 1970s concern over synthetic red dye in meat led to a resurgence of cochineal (labelled as carmine or carminic acid) in yogurt, popsicles, and Campari. But it's now linked to some allergies—a British medical report records the case of a man working with carminic acid in chorizo sausage who developed asthma. More than that, many deplore the duplicity of crushing pregnant females to produce rouge and lipstick: the website for PETA (People for the Ethical Treatment of Animals) says, "It takes a million corpses to make a kilogram of carminic acid."

EROTICISM, and the potential for violence, run through ink, letters, and their instruments. Sherkure, the beautiful woman in *My Name Is Red*, reflects on the connection between pens and oral sex: "I can't say I completely understood why Persian poets, who for centuries had likened that male tool to a reed pen, also compared the mouths of us women to inkwells ... was it the smallness of the mouth? The arcane silence of the inkwell?" The line of ink itself stirred the good juices: it was common to pun on *khatt*, which means both "script" and "down," the first black downy hairs of a growing beard on the cheeks and upper lip of a teenage boy. The daughter of the emperor Aurangzeb wrote,

> I have seen the library of the world page by page—
> I have seen your khatt and said, "That's the real purpose!"

The face was considered to be a "book of loveliness"; the Persian poet Saadi wrote,

> Due to the musk coloured khatt his cheek appeared like the
> book of beauty,
> for the commentator himself has written carefully these
> marginal notes.

And Ismail Bakhshi, a poet from the Indus Valley, explicitly links the sacred and the sensual, comparing the beloved to the Qur'an itself: "Your face is like a Qur'an copy without any mistake and flaw." Good penmanship was one of the qualifications for true beauty in a woman; many educated women were engaged in copying the holy book, and one eleventh-century calligrapher from Byzantium who was captured by Tunisian pirates became nursemaid to the prince's son, and produced one of the most beautiful Qur'ans. Indeed, the whole process of writing was eroticized—one female scribe inspired the vizier to the caliph al-Mu'tadid to exclaim,

> Her script is like the beauty of her form,
> her ink like the black of her hair,
> her paper like the skin of her face,
> her reed pen like the point of one of her fingers,
> her style like the enchantment of her eye,
> her knife like the flash of her glance,
> her cutting block (*miqatta*) like the heart of her lover.

As one commentator put it, "Arab calligraphers considered that their art was the geometry of the soul expressed through the body."

In Christian theology in the beginning was the word, but in Islam the first thing God created was the pen, with the divine light as ink. Abdelkebir Khatibi and Mohammed Sijelmassi in *The Splendor of Islamic Calligraphy* (1976) recount this myth of origin: "According to Abu al-Abbas Ahmed al Bhūni, letters arose from the light on the pen that inscribed the Grand Destiny on the Sacred Tablet…. After wandering through the universe, the light became transformed into the letter *alif* from which developed all the others."

The *alif* is the first letter of the Arabic alphabet.

It defines what comes after: the rounded parts of the rest of the characters are made to fit within a circle, of which the *alif* serves as the diameter. The *alif* is a symbol of the absolute oneness of God, yet poets often compared the straight, upright letter to the elegant, slender stature of the beloved. Hafiz (the fourteenth-century Persian poet whose love poetry is now pervasive in the West) writes,

> There is nothing on the tablet of my heart but the alif of the
> friend's stature
> What can I do? My teacher did not give me any other letter
> to remember

The height of the *alif* is determined by the particular style—in Nashkhi script it is five dots high, in Thuluth it is nine dots.

The size of the dot (a square formed by pressing the tip of the pen onto the paper) is in turn determined by dimensions and cut of the pen, as in this illustration of the letter *waw*:

The nib was divided into two lips that held the ink, and, Khatibi and Sijelmassi write, the stroke of the letters would vary according to "the pressure exerted by the fingers. This pressure had to be sufficiently delicate and precise to separate the two sides of the nib." Was it just me, I wondered, or did that sound erotic? The symbolism of any nib pen is of course both feminine and masculine. Art historian Annemarie Schimmel notes that the reed pen was often compared to the reed mat, on which the true lover would sleep—the lines of the hard mat would leave their marks on the body and thus "educate it in proper behaviour." Comparisons made between letters and parts of the human body and face were commonplace in Arabic, Persian,

Turkish, and Urdu poetry, and any educated person would know how to play with letter symbolism when writing elegant prose or poetry.

The letter *nun* was traditionally compared to the eyebrows of the beloved:

The *ayn* (an arch, and an invitation to share in Islam) was compared to the eyes:

The *waw* (which brings together all of Allah's creations) to the temple:

The *mim* (symbol for the Prophet Muhammad) to the mouth:

The *shin* (first letter of *Shari'a*, the Way) to the braided hair, as well as the tears that drop from the eyes of a lover:

The *lam* (symbol for the Angel Gabriel) to the long hair resting on the back of a woman:

To return to the letter *nun*, Sura 68 of the Qur'an begins, "*Nun, and By the Pen ...*"; here *nun* stands for the pen with which the Qur'an was written. The poet Rumi compares the letter to a Muslim kneeling in prayer (and he flips the metaphor, comparing the pious one to a "pen in prostration"), but by its shape the letter also signifies the primordial inkwell that belongs with that pen which has written on the Well-Preserved Tablet (the *lawh al-mahfūz*), which was the first thing to be created by God and which records all that is to be.

Ayaan Hirsi Ali, the controversial Somali-born political activist who was raised in Islam and is now an atheist and one of Islam's fiercest critics, writes in her memoir *Infidel* (2007) of a less-than-celestial tablet, and two episodes involving ink. The first is about going to Qur'an school: "I learned to mix ink from charcoal, water, and a little milk, and to write the Arabic alphabet on long wooden boards. I began learning the Qur'an, line by line, by heart. It was uplifting to be engaged in such an adult task." But, she says in the next paragraph, the atmosphere in the school was not uplifting. The kids were tough, and one girl, who was only about eight years old, "they called *kintirleey*, 'she with the clitoris.' … They spat on her and pinched her," and the teacher did nothing. The making of ink in the madrassah (which in Arabic simply means "school" but in English has come to signify institutions of Islamic teaching) is linked with the transition into the adult world, but also with sexual violence and betrayal by an adult authority figure. Hirsi Ali makes no explicit connection between the two; one just follows the other in her account.

But fifty pages later, ink comes up again. Her teacher does everything in the traditional Somali way, so Hirsi Ali "had to make ink before every lesson, scraping a piece of charcoal into powder with a broken piece of rough stone roofing tile, carefully dribbling milk and water onto the powder in a jam jar." One day her mother beat her because she hadn't finished the washing and cleaning, and "when the time came to make the ink I was already furious with the injustice of it all." She tells her sister they will lock themselves in the bathroom and refuse to work, but the mother goes out, leaving the door unlocked, and the *ma'alim* comes back with another man. Together they blindfold her and the *ma'alim* beats her with a sharp stick. Then he grabs her braided hair, yanks her head back, smashes it against the wall. "I distinctly heard a cracking noise," she writes. In her book the activity of traditional ink making is explicitly associated with betrayal, violence, and rebellion. It's not surprising that years later Hirsi Ali's protest takes the form of ink on the body.

In the short film *Submission*, written by Hirsi Ali and produced with Theo van Gogh in 2004, verses from the Qur'an relating to women (4:34, 2:222, and 24:2) were inked onto the body of an actress, who is shown naked under a translucent chador. At first the controversy was whether or not the makers were guilty of plagiarism:

one web journalist claimed they were influenced by the work of photographer and artist Shirin Neshat, who for her "Women of Allah" project had done a series of works with verses superimposed on women's bodies. But that issue soon proved incidental.

What happened next is well known: Mohammed Bouyeri, a Moroccan-Dutch Islamist, shot Theo van Gogh eight times with an automatic pistol, sliced his throat in an attempted decapitation, and stabbed a knife into his chest, with a five-page handwritten letter to Hirsi Ali attached. Some French philosopher said that all disgust is tinged with desire, and it seemed to me that the sacred is always tinged with violence, the erotic suffused with destructive power. The film, now inseparable from its aftermath, fascinates us because it brings together the sacred, the erotic, and the violent in an extreme conjunction.

WHAT UTHMAN PRODUCED in that first Qur'an, as opposed to the ornate Qur'ans in Istanbul's Topkapi museum, was the *rasm* of the text—the consonantal structure, without the dots and squiggles (the diacritical marks that distinguish different values or sounds for the same letter, as in the French *é* and *è*) and without vowel signs, which were incorporated later.

Arabic script with diacritical marks, Metropolitan Museum of Art, New York. Photo courtesy of the author.

The look of the letters is clean, sleek, and unambiguous, but that's because they're just the scaffold. It's as if instead of the word *Bible* you had *bbl*. This could lead to variant readings:

bible
babel
bauble

The example from the translations of the *Rubaiyat* in which "A book of verses … a loaf of bread, a jug of wine and Thou" can be read as "Roast meat … a loaf of bread, a jug of wine and Thous" is whimsical, but variant readings can shape, and divide, a whole culture.

Majid Fakhry, of New York University, translates The Women 4:34, the famous verse about the obligations of men toward women, as follows:

> Men are in charge of women, because Allah has made some of them excel the others, and because they spend some of their wealth.… And for those [women] that you fear might rebel, admonish them and abandon them in their beds and beat them.

But Ahmed Ali, in the Princeton edition of the Qur'an, renders it this way:

> Men are the support of women as God gives some more means than others, and because they spend wealth (to provide for them).… As for women you feel are averse, talk to them suasively; then leave them alone in bed (without molesting them) and go to bed with them (when they are willing).

Surely one of these translators is twisting the text to serve his own ends? Not at all, says Reza Aslan, in *No god but God*: "Because of the variability of the Arabic language, both of these translations are grammatically, syntactically, and definitionally correct." The final word in the verse, *adribuhunna*, which Fakhry has rendered as "beat them," can equally mean "turn away from them," "go along with them," and even "have consensual intercourse with them."

So different versions of the Qur'an had different readings, even though they might have the same *rasm*. What would the Prophet have wanted? That's not an easy question to answer. In an account of the Prophet settling a dispute over readings, he is said to have declared, "Each of your readings is correct. The Qur'an was revealed in seven forms, so recite whichever is easiest." So although Uthman ordered all other versions to be destroyed, it has been argued that the script without diacritical marks was a deliberate way of accommodating the seven readings sanctioned by the Prophet.

II

Yours is the most generous Lord
Who taught men by the pen
That which they did not know
—QUR'AN, THE CLOT 96:3–5

The amorphousness of Central Asia continued to baffle me—I wasn't even certain what I'd seen in Tashkent. I learned that only one-third of the original Qur'an of Uthman still exists and that dummy pages made of paper, not gazelle, have been substituted for the missing leaves. Some of those pages have come up for sale at Christie's and Sotheby's, and other Qur'ans have surfaced claiming to be stained with the blood of Uthman. Also, it appeared that even from the beginning there were other Qur'ans contending with Uthman's.

Back in Edmonton I brought up the problems over brunch at the Glenora Bistro with two Muslim friends, Tim, an Irish-descent Canadian, and Samina, a Montreal Pakistani woman for whom he had converted to Islam when they were both in medical school. They seemed very secular. "It wasn't hard giving up drinking," he said. "What was hard was bacon." This seemed to be the biggest issue. Samina told me how their toddler took the song, "Who built the ark? Noah! Noah!" and turned it into "Who built the ark? No one! No one!" Yet she became more serious when I started talking about the Qur'an. She said, "When there were funerals and we all read through the whole Qur'an, we kids always asked for chapter thirty because it was easy to read and we could get through it fast. 'I want thirty!'" She smiled at the memory.

I returned to my question. "There were variants—Uthman had assembled his version, but over in Kufa a man named Mas'ud had a Qur'an that didn't have the first chapter or the last two."

"Really? But those are the first ones you learn! Because they're short and they're easy." Samina laughed. "And that last one is the one that precedes everything ..." She trailed off. "Can we talk about something less depressing?"

It was part of her youth, and something absolute. Though I'd read the Qur'an from beginning to end, I'd been reading as a researcher, obsessed with questions of textual scholarship yet with no feeling for the text itself. Now for the first time I had the sense of the Qur'an as a living document, the spirit of a culture, and something that infused the family life of the friends beside me.

On the way back from Uzbekistan I'd made a side trip to Istanbul. I took the tram out to the end of the line and caught a taxi to Emirgan, now part of greater Istanbul but once a separate fishing village on the Bosphorus. The road passes clapboard houses that remind a Canadian of St. John's, Newfoundland, and *yalis*, the dark wooden villas that the Istanbul elite retreated to in the heat of the summer. Now instead of private boat docks there are trendy cafés and fine restaurants by the water, and mansions on the hills, and among them, on green rolling grounds, the Sakip Sabanci Museum. Their exhibition, "Transcending Borders with Brush and Pen," drew together works from Japan, China, Europe, and the Ottoman Empire.

They had pages from a hand-lettered Bible—brown ink in a Gothic hand, double columns, and chapter headings in red and blue with a two-line initial letter in raised gold, red, and blue— the work of nameless scribes who had influenced Gutenberg. They also had a manuscript by Wang Xizhi commemorating the banquet at the Orchid Pavilion, where the literati had drunk wine from goblets as they floated down the stream, and in a room nearby, resplendent in black, gold, and red, was a page from a Qur'an by Seyh Hamdullah, one of the most celebrated calligraphers in the Islamic world. The stated aim of the exhibition was to affirm that writing was more than a mere vehicle for ideas, that it embodied the internal world of individual human beings and represented the concentrated expression of a culture. The objects conveyed the flow of ink through cultures, linking them on a material level, gesturing toward the spirit that united them.

Later I lay in my hotel room with its view of the famous Blue Mosque and in a dream-doze listened to the call to prayer. To a Westerner the cry is initially strange and annoying, like an alarm clock going off at odd hours. Then you find you're unconsciously waiting for it, welcoming it. It transports you, feels like a call to leave this world and enter another one. So too does the calligraphy beckon, gesture, invite the transformation from ink to breath. It is a commonplace that the two things that unite Islam, across cultures and countries, are calligraphy and the call to worship, the *adhan*. Professor Seyyed Hossein Nasr of George Washington University, writing on the place of oral transmission in Islamic education, notes that in Islam there is a tradition of reading "the white parts," what in English we call "reading between the lines." The idea wasn't new to me—Virginia Woolf wrote that meaning lies just on the far side of language—but in Islam the auditory aspect is an essential part of interpretation. Uzma Mirza, an artist and architect (who spent part of her youth in Cape Breton), writes of how reading the Qur'an is to enter a "sonorous space" in order "to know the One, from which she [the reader] is a reverberation." The text exists beyond the book, and in talking with Samina I began to see how this was so, how for her the spirit of the Qur'an was embodied in its practice.

Traditions of Islam that I thought foreign turned out to permeate my own culture. I knew the concept of the *lawh al-mahfūz*, the Well-Preserved Tablet, on which the destinies of men have been engraved since the beginning of time, and of the Primordial Pen that sets down everything that is to happen. But I learned that there are echoes of the miraculous pen in the travels of Marco Polo and in Boccaccio's *Decameron*, where Friar Cipollo (Brother Onion) exhibits the quill feather of the angel Gabriel, and that the concept lies behind the phrase "it is written" (*maktub*), which expresses the inevitability of fate.

More intriguing, for anyone interested in the origin of the novel form, Miguel Cervantes's 1605 *Don Quixote* ends with a pen hanging from a wire; the pen declares, "For me alone Don Quixote was born and I for him. His was the power of action, mine of writing." The critic Luce López-Baralt argues that this pen of Cide Hamete Benengeli (the Moorish pseudo author of the novel) resembles the Supreme Pen, the *al-qalam al-a'la*, of the Qur'an (68:1), which writes without any human intervention. Cervantes, who'd been imprisoned in Algiers and who was acquainted with Moriscos (Moors who converted to Christianity after the expulsion of the Muslims in 1492) in Spain, would have been familiar with Islamic traditions. Reading back to the Qur'an through *Don Quixote*, López-Baralt observes that the *Nun* passage from the Qur'an—"By the Pen and what they [the angels] write you are not mad: thanks to the favour of your Lord!"—suggests the demented Don from La Mancha (which means "the stain" and by extension "ink stain"). And so just as ink filtered from the Middle East through to Europe, the tradition of the pen was modified and absorbed into our own literary tradition, with the book that defines a new genre in Western literature—the novel.

Connections were cropping up everywhere. In her essay "The Pen and the Inkpot: A Muslim Woman's Spiritual Art," Uzma Mirza reiterates that in Islam, calligraphy is seen as the human emulation of the divine act of creation—each letter has a science, melody, and personality, and corresponds to specific qualities of Allah. In a surprising move, Mirza quotes Maria Montessori, who discovered that children learn to write several months before they learn to read: "Where did it come from?" Montessori asks. "No one enforced it; and what is more, no one could have obtained it by external means. Had these children maybe found the orbit of their cycle like the stars that circle unwearying and which without departing from their order, shine through eternity?" The Celestial Pen.

At Pages bookstore in Calgary, Hsing found another book for me, inexpensive but exquisite: the Porcupine's Quill edition of *The Essential P.K. Page*. I'd forgotten that this very Canadian poet, who'd lived in Red Deer, Calgary, Winnipeg, Montreal, and finally Victoria, had studied Sufi mysticism. In "The Filled Pen" she writes of how the pen and ink both unlock expression and grant access to the animating principle of the universe: "the delicate nib" not only releases the "huge revolving world," but reveals "the rarely glimpsed bright face / behind the apparency of things." The Samarkand Codex, the thing furthest afield from me in both geography and culture, had led home, casting my friends, my national literature, and my research in a new light.

The ballpoint and the printing press celebrated the democratic utility of ink. Although Gutenberg's first big project was a Bible, he was already busy printing Latin grammars; it was the spirit of capitalism that drove him, not the divine. And the spirit of Chinese ink, with its emphasis on character, is philosophical rather than religious. It was Islam that transported ink to the spiritual realm. Although an element of spirituality has always been present in the ink of holy books, from the Dead Sea Scrolls to the Bible to the Torah (with its 304,805 letters, each of which must be perfect, written with a reed pen in gallnut ink), only Islam makes the celestial pen part of its creation myth, of the world, inscribed before being, in divine ink.

In a sense, the Qur'an is the archetype of writing. Written in ink that will last more than a thousand years, the physical book, the *mus'haf,* is still only a place holder for the true Qur'an, the Preserved Tablet that is with God. Ink is always the trace of thought, of spirit in the largest sense of the term. The distinction between the letter of the law and the spirit of the law is basic, and all writing defines a negative space, a space not of absence but of plenitude. The work you're reading is simply black marks on a page. The text that derives from it takes shape in the mind. Thus all texts are shaped by experience and context, and are always different, even for the same reader: it is a truism that you can never dip into the same book twice. The ink gestures toward that which transcends it.

PART IV

A *Renaissance* OF

ink

Shadows, he thought, are like ink.
They are shady and shifty and mysterious.

—BRIANNE FARLEY, *IKE'S INCREDIBLE INK*

Pen, notebook, and coffee, Budapest café.
Photo courtesy of the author.

This Twittering World

I

"Dickens's quill has been splodging horribly and scratching and is therefore being retired this evening." Thus declared Colin Thubron, eminent travel writer and head of the Royal Society of Literature, on March 18, 2013. For decades the Society's newly elected members have signed the roll book either with Byron's pen (still in good shape) or Dickens's now-splodgy quill, which has been replaced with T.S. Eliot's fountain pen. It's a Waterman 16 PSF (Pocket Self Filler) with gold barrel bands, engraved T.S.E., given to Eliot by his mother and used by him his whole life. Colin Thubron—who likes the "immediate and personal act" of composing with a pen as well as the "discipline" of transferring his text to the screen—told me that the Society uses Waterman's black ink in the pen and that it writes very smoothly. Eliot too hovered between technologies: "Composing on the typewriter, I find that I am sloughing off all my long sentences which I used to dote upon. Short, staccato, like modern French prose. The typewriter makes for lucidity, but I am not sure that it encourages subtlety."

Words change by not changing as the world changes. In the 1930s Eliot wrote "Not here / Not here the darkness, in this twittering world"; even then he saw our reluctance to descend into the fruitful darkness of solitude. In my old edition of *Four Quartets*, studded with pencilled banalities, the passage is unmarked, but it leaps out now (and appears on innumerable Twitter feeds). I'd marked the passages on language, liked how they unfolded, overlapping:

> Words after speech, reach
> Into the silence. Only by the form, the pattern
> Can words or music reach
> The stillness, as a Chinese jar still
> Moves perpetually in its stillness.

The "reach" into "stillness" now defined calligraphy for me, its beauty and accomplishment, but Eliot wrote most feelingly about the failures of language, about how words "slip, slide, perish, / Decay with imprecision, will not stay in place, / Will not stay still." Any reader who's seen their hard drive go south and known their words have perished will identify in a way Eliot never could have imagined. Anyone who's had their point mangled on social media will feel acutely how words will not stay in place, will not stay still.

When I was in my late twenties, I wondered at myself liking this old-man poetry. What brought me back to Eliot was the half-remembered "the end of all our exploring / Will be to arrive where we started / And know the place for the first time." A common enough sentiment, but he puts it better than most, and I'd just run up against it. I'd proved to my own satisfaction that ink meant nothing to the West. It was a utilitarian substance, a craft not an art, its only romance the perfection of a throwaway object. As an undergraduate I worked on the university grounds crew, and one year I was paired with an engineering student from Iran, digging holes to transplant shrubs. As we dug we talked about poetry. I was surprised: I thought engineers hated poetry. He said, "In my country poetry is everywhere—on the radio, in the street." Here in Canada poetry wasn't part of daily life; it was ghettoed in small readings, small sections of the bookstore. The letters of the Western alphabet can be beautiful, but they don't sway like the "tense quivering" of the beloved's calligraphy in *My Name Is Red*. There is no Western tradition that merges the sensual and the sacred in the shapes of the letters. Part of the power of Theo van Gogh and Hirsi Ali's film *Submission* derives from the explosive conjunction of sensual letters of the sacred text written on the naked body of a woman. I could not imagine someone being murdered for painting the "Song of Solomon" on a *Playboy* centrefold.

And as for the ink itself, none of our poets talks about "ink like the soaring wings of the raven." European ink recipes cannot compare to those of ibn Badis, who calls for ink from anemones, stalks of roses, grand basil, indigo, sumac, bee honey, and mica, or scented with the pendulous white flowers of the styrax tree. Although there's a lip-smacking sensuality in the way the rollers of a printing press take up the ink—and doubtless someone has thought of smearing it on the body of the beloved and rolling him or her across the page—in Western culture there is no prominent link between ink and the erotic. (If there were, I'm sure British film director Peter Greenaway would have capitalized on it. In his 1996 film *The Pillow Book*, a Japanese model, played by Vivian Wu, derives sexual pleasure from being written upon and writing on the bodies of her lovers, one of whom she rejects because his skin is no good for calligraphy: it makes the ink run. She finds satisfaction with a British translator, Ewan McGregor, whose skin takes the ink so well that eventually a book is made from it. The metaphoric implications of McGregor's nationality, and of the publisher who exhumes his corpse to make a book, remain tantalizingly obscure but open up questions of ink and cultural difference.) Our tradition was all about function, and now ink was effectively dead.

However, I hadn't been paying attention. Was it in Herman Hesse's *Siddhartha* that a character says just because you've departed from the way doesn't mean it doesn't exist anymore? I came across Charlie Paul's 2012 documentary about Ralph Steadman, *For No Good Reason*, in which Johnny Depp interviews the gonzo cartoonist/artist at his studio. The most dramatic sequences show Steadman splattering ink onto paper and then making a detailed cartoon—not an abstract design—out of the ink splash, the image emerging from the ink itself. I encountered the children's books of Steve Light, who sketches in pencil and then inks the drawings with his Montblanc. Pens figure in the books themselves—in *Zephyr Takes Flight* (2012) the young heroine inherits her grandfather's Conklin fountain pen, while *Have You Seen My Dragon?* (2014) includes a fountain-pen store on the back cover, and, inside, an illustration of the author working with his pen. In Brianne Farley's *Ike's Incredible Ink* (2013) the ink itself is the focus: Ike can't settle to write his story until he has the right ink, and so he decides to make his own, which involves prying loose a shadow, collecting matter from the dark side of the moon, mashing, crushing, and generally making a mess.

Further, writing was back in the news. *Pen World*, the glossy mainstream magazine, broke the story to the pen community at large of how in 2011 scientists from the School of Mechanical and Aerospace Engineering at Seoul University in Korea and from the School of Engineering and Applied Sciences, Department of Physics, Harvard University had established—for the first time ever—"how liquids spread on a rough substrate (paper) from a moving source (pen)," in other words, how ink flows. The article soon leaves the lay reader behind, plunging into the varying velocities "owing to viscous shear" at the back of the "parabolic front" of the line of ink as it fills "the gaps of the forest of micropillars" on the paper ("balancing the driving force of spreading in radial direction $y(f—1)r\Delta\theta$ with the resisting force μ …"). At last, the mystery revealed—though I couldn't really understand it. What I found most remarkable was the authors' concluding hope that their study would have the effect of "perhaps even rejuvenating the ink-pen in another guise"—an implicit belief that even in an age in which new digital devices are launched every six months, pens and ink are worthy of development.

Laurence Bich had said that in medieval times people wrote their names once or twice in their life. A recent news article reported that many high school students can't even do that—can't sign their own names. Should we be appalled? (Pre-literate kings used to make the sign of the cross next to where the clerk had written their names, thus "signing.") An Associated Press story reported that when Barack Obama promoted his White House chief of staff Jack Lew to treasury secretary, the president joked that he nearly rescinded the appointment because Lew's signature, which will appear on the dollar bill, was just a J followed by seven loopy scribbles. A contrite Lew promised to work to make at least one letter legible, "in order not to debase our currency." Cursive, even your own signature, is a joke. Hamlet wanted to affect bad handwriting, but even he would probably balk at signing his name "Hooooooo." Has the link between handwriting and character been severed?

Apparently not. During one of actor Lindsay Lohan's court appearances, handwriting experts analyzed her blocky courtroom scribbling—"projecting a false image" and "crossing boundaries," concluded two pundits on hollywoodlife.com. And Sandra Folk, writing in the *Financial Post*, says, "Don't write off the power of the handwritten note": after a job interview she suggests an email acknowledgment and a handwritten letter, because the latter "demonstrates your thoroughness and attention to detail" and "makes you stand out from the mass of applicants, your competition."

According to scientists, whether you're a preschooler or a boomer trying to ward off Alzheimer's, forming letters by hand is good cognitive exercise. Virginia Berninger, professor of educational psychology at the University of Washington, says that because handwriting requires sequential strokes rather than touching a key to select a whole letter, it activates regions involved in thinking, language, and working memory. One of her studies demonstrated that in grades two, four, and six, children wrote more words faster and expressed more ideas when writing essays by hand than when using a keyboard. The neuroscience supports her. In an essay on "the haptics of writing," a pair of scientists from Norway and France write about their findings on the physicality of writing. The field of haptics (from the Greek *haptikos*, to touch) deals with tactile perception, and relates to the larger field of embodied cognition. As babies, we first explore the world through touch, but touch is important for adults, too; it's the most primordial of the senses, and it informs many of our expressions for comprehension—we "grasp a concept," learn to "handle a situation," try to "get hold of someone."

Handwriting is bound to the physical body and creates a different kind of concentration, activates different neural pathways. On a keyboard there's a "decoupling"—not only are you working with two hands, your attention is split into two distinct fields: the visual field of the screen and the motor field of the keyboard. With a pen you use one hand, and your visual attention is concentrated at the tip of the pen, where the ink flows. Further, whereas on a keyboard the physical action is the same for every letter, in hand-writing you have to produce a different graphic shape for each one. Brain imaging studies show that the specific movements involved in writing support the visual recognition of the letters, and may be a factor in learning to read. So when writers talk of the "kinetic melody" of handwriting and the "visual melody" of the text, of the "sculptural pleasure" and the sense of "craft" that comes with using a pen, they may sound romantic or nostalgic, but their words point to a hot area of scientific research: the materiality of learning, the corporeal nature of knowledge.

The constraints of handwriting in fact aid composition. Canadian poet Tim Bowling shocked an audience of writing students by telling them that he never revises. Then he elaborated: he composes on long walks and, because he had a schoolteacher who stressed neatness and good penmanship, he carefully writes out the draft. No strikeouts allowed. Much revision takes place in his arm, between his brain and the point where the words flow into ink. This technique works not just for poets walking in the ravine, but for journalists working to tight deadlines. Handwriting is a way of knowing.

Marshall McLuhan spoke of all media as "the extensions of man"; pens and pencils are extensions of our fingers. David Chandler, in his much-cited article "The Phenomenology of Writing by Hand," speculates that there may not be the same sense of extension with typewriters and computers because you no longer "touch" your texts. He cites D. Kruger's *Introduction to Phenomenological Psychology*: "One's tactile experience locates itself in the tip of the pen—one feels it scraping or gliding across the paper. The living body therefore goes beyond the boundary of the skin and incorporates the pen as part of its structure." Robert Metcalfe, a friend of mine at university, now a respectable Montreal lawyer, used to refer to the phenomenon of the "magic pen": he would sit down to write with no ideas in his head, but then they would come—"It's like they're in the pen itself," he said, looking wonderingly at the tip. One of Chandler's interviewees turned this around, saying he himself becomes an extension of the pen—"It is the act of writing that produces the discoveries ... words flow from a pen, not from a mind.... Consciousness is focused in the point of the pen."

Experiences I dismissed as foolish or idiosyncratic turned out to have a grounding in wider intellectual inquiry. First thing in the morning, along with a cup of coffee, I want the feel of the pen on the page, a pleasure I thought facile until I found it validated by Chandler, and by Frank R. Wilson, who's written a whole book on the subject, *The Hand: How Its Use Shapes the Brain, Language, and Human Culture*. Wilson has technical chapters on the physiology of the hand but also fascinating case studies of puppeteers, jugglers, cooks, surgeons, hot rod mechanics, and matadors. He takes as his mantra a line from the Robertson Davies novel *What's Bred in the Bone*: "The hand speaks to the brain as surely as the brain speaks to the hand."

In a *Guardian* article on British novelist Alex Preston, who's composing his next novel entirely in longhand, the interviewer himself says that using pen and notebook keeps him in touch with the craft of writing. "It's a deep-felt, uninterrupted connection between thought and language which technology seems to short circuit." Canadian journalist Andrew Coyne agrees: "Text on a computer is infinitely corrigible: We commit to nothing, either in words or sentence structure.... Handwriting, to the contrary, forces us to make an investment. It inclines us thus to compose the sentence in our heads first—and the sort of sentence you can compose and keep in your head is likely to be shorter and clearer than otherwise." Military historian Nathan Greenfield says part of the joy of using a fountain pen is that it's a beautiful object, but also that it gives him a different feel for language. He composes on the computer, yet when he's stuck or sketching a new idea he turns to the pen. There's another factor, too, although he's diffident about admitting it. In conversation he said, "This is where the military historian turns into Romantic idiot, but when things are going well I feel that the fountain pen connects me to a tradition … the nib itself is an historical object in a way that the Bic pen isn't."

When you get up to speed, of course there's no substitute for a good keyboard, but cognitive dissonance comes with the convenience. The eagerly intrusive software will finish your words for you, but when it turns "Amin Maalouf" into "Amin Meatloaf" (as it did for me) it derails your thought process (What *would* he do with "Paradise by the Dashboard Light," I wondered). Email now upsells your memos—"Do you also want to send this to X?" A pen isn't going to ask "Do you want to write a note to your mother as well?" even if you know you should. It's only now that we spend so much time on screens that we appreciate the calm of the page. With pen and ink you're your own software. And, an added bonus, there's no metadata; no one tracks your notebooks.

If the pen stimulates creativity, the computer, at least in the early stages, may stunt it. In a Romantic simile, David Chandler writes of how for some writers words "spill out like hot lava … gradually cooling and setting" into texts; he speculates that the fondness for the pen may be related to "a peripheral awareness of the ink drying on the paper," which mirrors one's own thoughts "passing from wetness to setness" (and also mirrors the ghostly way gallnut ink appears on the page). Chandler suggests that the word processor may seem to make writing cool off too fast. In fact, some writers even find ink too solidifying. Hemingway said, "You have to work over what you write. If you use a pencil … it keeps it fluid longer so that you can improve it easier"; and Henry Petroski in his 1989 book *The Pencil* suggests that it's "the ephemeral medium of thinkers, planners, drafters and engineers, the medium to be erased, revised, smudged, obliterated, lost— or inked over," contrasting it with ink, which "signifies finality." And, even worse, the laser printer makes the text look like a book rather than a manuscript in progress, allowing the delusion that the work is done.

Iris Murdoch saw this back in the early 1980s: "The word processor is … a glass square which separates one from one's thoughts and gives them a premature air of completeness." On our computers we look at text through a window, which limits the number of lines that are immediately visible, and which leaves you feeling lost in space when revising larger structural blocks. The size and clarity of the windows have improved since Murdoch's era, but we're still on the outside looking in. Her observation echoes the famous passage from Robert Pirsig's 1974 *Zen and the Art of Motorcycle Maintenance*: "In a car you're always in a compartment, and because you're used to it you don't realize that through that car window everything you see is just more TV…. On a cycle the frame is gone. You're completely in contact with it all. You're in the scene, not just watching it anymore, and the sense of presence is overwhelming." With the pen on the page you're unbounded, outside the frame; you can scribble above, below, make drawings. Tablet devices enable you do that, but they still lack the connection and "presence" Pirsig talks about. LiveScribe has tapped into this desire with their Echo pen: it's also a recorder and a camera, and it can store audio files and convert notebook pages to pdf, yet it remains, on its most basic level, a ballpoint pen.

Riding a motorcycle and writing may seem streets apart, but in both you're out there with naked metal doing the thing itself, absorbed in the flow. Dante Del Vecchio, founder of Visconti pens in Italy, races a motorcycle, and Philip Wang, the young, hip Taiwanese-Californian inventor of the TWSBI fountain pen, races a Ducati Monster. Each of their pens comes with a little wrench so that the owner can dismantle it, a legacy, Philip told me, of his father's love of working on motorcycles.

Motorcyclists ride for the ride, pen users write to write; nobody keyboards to keyboard. Motorcycle-culture theorist Steven Alford says, "On a motorcycle you don't say, 'Let's go to X,' you say, 'Let's go for a ride.'" You probably have a tentative destination, some notion of a route, but it's not fixed and can be changed at any moment. So too with the pen on the open page. You're going for the ride, you're not committed to a destination, your route will not unfold in straight lines. If you write a wrong letter you just groove it over; when you want to make a change you can nip back, strike out, add a phrase; and through it all you won't have a winking cursor perpetually asking "What's next?" You can pause, enjoy the stillness of the page. For Alford, "the essence of motorcycling is ... the effacement of the self in an experience of 'flow,'" a state familiar to mystics and practitioners of extreme sports in which, completely absorbed in the action, you lose all sense of time and selfhood. The rider becomes "a combination of the environment, the motorcycle, and his body. He's not travelling, he's enacting movement." If we replace "motorcycling" with "writing," and "motorcycle" with "pen," we get: "The essence of writing is the effacement of the self in an experience of flow.... You become a combination of the environment, the pen, and your body." Which leads to "You're not transcribing, you're enacting thought."

But there's more.

II

Roland Barthes in his 1973 *Le Monde* interview confessed, "I have an almost obsessive relation to writing instruments. I often switch from one pen to another just for the pleasure of it. I try out new ones. I have far too many pens—I don't know what to do with all of them! And yet, as soon as I see a new one, I start craving it. I cannot keep myself from buying them." He's profligate in his purchases, though he has his standards: he says he's tried everything, "except Bics, with which I feel absolutely no affinity. I would even say, a bit nastily, that there is a 'Bic style,' which is really just for churning out copy, writing that merely transcribes thought…. In the end, I always return to fine fountain pens. The essential thing is that they can produce that soft, smooth writing I absolutely require." There is a promiscuity to pen buying. One pen manufacturer speaks of pen nibs lasting eighty years, but that ignores the reality of the pen enthusiast. Why a store filled with so many pens, so many inks? So you can find just the right pen, buy a gallon of ink, and keep it until you die? I think not. Most pen owners would acknowledge Barthes's "as soon as I see a new one, I start craving it." Utility, thrift, the environment—they have nothing to do with it.

Just as the theme of the sexual runs through much of Barthes's writing ("jouissance," which carries the sense of both bliss and orgasm, is central to his 1973 essay, *The Pleasure of the Text*), the act of writing is itself both sensuous and sensual. *Sensuous* was a word invented by Milton to avoid the sexual associations of *sensual.* Strictly speaking, it means "relating to the senses rather than the intellect." Unfortunately, in common usage the meanings have become blurred, especially after the 1969 publication of the sex-help book *The Sensuous Woman*, but the distinction is worth preserving. Writing is obviously sensuous—my department chair tells me she loves crossing items off her to-do list, so much so that if she does something that's not on the list she'll write it down just so she can slash it off. Ticking the box on her computer doesn't do it. Without the tactile engagement there's no visceral thrill—indeed, the verb *write* derives from a Teutonic root signifying to scratch or tear. But writing is also less obviously sensual, which, the *Oxford English Dictionary* warns us, "involves the appetites or desires," is "carnal, fleshly," and can be "lewd, depraved." It took me years to realize it, but like Barthes, I "absolutely require" writing—not to compose, not to write in the literary sense, but to move a pen across the page. When those who love to write haven't had any writing in a long time they get restless, irritable. Their muscles twitch, their scalp feels scritchy, and they can't wait to get to it. When they do they can hardly hold back, and the words gush forth unconsidered—the forming of the letters, the feel of the nib gliding across the paper, the flow of the ink out onto the sheet, that's what one wants. Calmness follows, thinking slows, the grip relaxes. They begin to think in sentences again, the pleasure of the page now mingled with the impress of thought.

In her essay "Eternal Ink," Anne Fadiman writes about the sensuous joys of writing with a fountain pen. She hates those "disposable models that proclaim, 'Don't get too attached. I'm only a one-night stand.'" For Fadiman, "Like a dog that needs to circle three times before settling down to sleep, I could not write an opening sentence until I had uncapped the bottle of India ink, inhaled the narcotic fragrance of carbon soot and resin, dipped the nib, and pumped the plunger—one, two, three, four, five." She loves stories like the one about Sir Walter Scott, who, while out hunting and without a pen, thought of a line and "shot a crow, plucked a feather, sharpened the tip, dipped it in crow's blood, and captured the sentence." She's not a Luddite, and is surprised by how much she likes her computer, "but," she declares, "I will never love it."

Can you love a computer? Teenagers save their old cell phones in shoeboxes, a friend told me, and I admitted I could never recycle the Panasonic Toughbook I took to Texas on the back of a motorcycle and carried across India in a knapsack. One can love both. Peter, the owner of Laywine's pen shop in Toronto, said, "I ask people, Do you have a bike? Do you have a car? Would you sell your bike?" It's a different experience; one is not a replacement for the other. And as with bikes, ink itself is becoming a fetish object. Peter showed me his most expensive bottle—I quickly computed that its ink cost the equivalent of fifteen pints of Guinness. Yet beer was obviously the wrong point of comparison. The packaging was beautiful, the bottle itself exquisite: the ink was just slightly cheaper than the Chanel No. 5 Eau de Parfum available in the boutique down the block. Ink has entered the domain of exclusive accessories.

*Waterman's blue-black, in the bottle that tilts so that you can
fill easily even when the level is low. Photo courtesy of the author.*

The Community of Ink

I

I wanted the right blue.

Not too bright, not in fact too blue, a blue-black. But not black with a hint of blue, and certainly not one of those blue-blacks that look great out of the bottle but dry to a blue-grey. I discovered an ink company called Private Reserve that made several dark blues I liked. I bought a bottle, and after a few fillings bought the even darker blue. Then, after a few fillings more, bought the slightly lighter blue. Three blues. When the time came to refill my pen I would, I won't say "agonize," but certainly "consider" which ink to use. I knew no one who did this. And it didn't stop there. I wanted to buy more ink. I was hiding bottles from my family. Ink was becoming a secret vice.

"It's the internet that changed everything," said James Leech, the owner of the Stylus pen shop in Edmonton, where in just a few years the shelf of inks had changed to a wall of ink. "It brought people together. People thought they were the only ones; maybe they knew of one neighbour up the street. Now they can connect all over the world. There's a community." We were talking about the resurgence of interest in fountain pens, and the word *community* kept coming up. If you were an enthusiast, the Fountain Pen Hospital in New York City had been operating since 1946, but before the internet, unless you were in lower Manhattan on a weekday, you'd be unlikely to stumble across it. Two decades ago fountain pens were something to be given as presents. They would appear in gift shops at Christmas, in the university bookstore just before graduation time, and for inks you had a choice of blue, black, or blue-black. Waterman or Parker.

Now James has inks from fifteen or sixteen manufacturers on his wall, with as many as a dozen colours from one manufacturer. "Private Reserve has fifty colours, Diamine now is up to one hundred," he said. "We can't stock them all." You don't buy blue anymore, you explore a range of micro-tones. Private Reserve offers American Blue, Cosmic Cobalt, Daphne Blue, Ebony Blue, Supershow Blue, Lake Placid Blue, Naples Blue, Sonic Blue, Tropical Blue, Midnight Blues, Electric D.C. Blue, and Black Magic Blue. If you're feeling continental you can go to J. Herbin (but pronounce it "ash-air-BAN," not "jay-HERB-in" as I first did) for a choice of Bleu Nuit, Bleu Myostis, Éclat de Saphir, Bleu Pervenche, or Bleu Azur. A historian by training, James had always loved pens, and he figured there was a community of pen enthusiasts, and beyond that a community of people who were interested, who would be drawn in, who wanted to be educated. "In the old days your pen was something you'd hand down with your watch. Nobody ever thought of throwing a pen out. Bic changed the perception of a pen into something cheap and disposable. It's what pen dealers are trying to counter today." He's done well; his shop has developed a cult following. "We've been lucky," he said. "A lot of shops in the American northeast have been closing. If Goulet Pens had a bricks-and-mortar shop he couldn't make it."

In fact I had a bag of samples on the way from Goulet Pens, and so the next day I talked on the phone with Brian Goulet in Virginia. I'd already established that he was a distant relation of Canadian crooner Robert Goulet, and I knew what he looked like from his informational videos on pens: early thirties, heavy-set with close-cropped hair, thick eyebrows, and beefy forearms, always wears T-shirts; he looks like a football player, not a pen nerd. "We wouldn't exist except for the internet," he said. "We still have no paid advertising, and our customer base has been created through blogging, Twitter, Facebook, the Fountain Pen Network. It's grass-roots." It has grown along with social media. When Brian graduated from college in 2006, "blogging was comparatively new, YouTube wasn't so common, and I had no idea how to make videos." Now his YouTube video channel, Ink Nouveau, is a go-to source for information about pens; his videos have already racked up 35,000 views. "I know that's not much compared to videos of a cat doing tricks," he said. "But it shows people are interested." His crisp five- to ten-minute talks in his "Fountain Pen 101" series draw grateful comments; people want the information and he has a good sense of what the customer needs, perhaps because he himself came late to the fountain pen world. His "conversion experience," as he calls it, came in his twenties.

"I started out as a pen maker, turning wooden pens on a lathe, making ballpoints. I was good at making pens but not very good at selling them." They'd be corporate gifts, a one-time purchase; Brian wanted a sustainable business, and "there was no community around the craft." There was that word *community* again. "I wasn't even really aware of fountain pens," he said. "But I attended the Pen Show in Washington, DC, in 2009, and I went home that night and joined the Fountain Pen Network" (the major online discussion group for pen users). He was excited, determined to make his own fountain pens. "But they weren't very good. They were made of wood and very heavy, very thick." So for the first year he made his focus ink, along with some paper, because he saw a gap in the market, a need in the community.

"What's great about writing by hand is that you can customize the experience—the paper, the pen, the ink." For Brian, ink was always the essential element. He considered what colour he'd use that day, what suited his mood. He loved the feel of the ink going onto the page. "One of the tables at the DC show had sample swabs, and the ink looked great, so I bought a bottle and took it home and it didn't look the same at all. It looked different on my usual paper, looked different from a pen rather than applied with a Q-Tip, as they do at pen shows. I wanted to try different inks, to write reviews, but it was expensive to buy whole bottles, and I wished I could get samples. I figured if I wanted this then other people must want it too—and Fountain Pen Network had this ink-sample exchange. That was the inspiration. So after nine months we started selling samples. We had thirty-eight at first and now we have over six hundred."

I knew. I'd ordered eight a few months before, and had just ordered another twenty (and made an impulse buy of a flat-black Lamy pen—hardly expensive, and although I already had five Lamys I didn't have one in that colour; the same argument, I recognized, that Hsing had just used to justify more shoes). I told him it was the samples that had brought me to his site, led by word of mouth, and although it was wonderful being able to buy samples like buying a single song on iTunes, it was dangerously addictive. *Only a buck fifty*, you think as you click on another colour. Brian laughed. "Yes, people call me a pusher."

Did he think the pen would endure?

"The fountain pen isn't required to function in life; it's a choice," he said. He regrets that schools have stopped requiring cursive writing, but he doesn't think that's going to stop people from buying pens and ink. It's precisely because it's not something enforced that makes it attractive. "Some are drawn to it because it's eco-friendly: you're reusing your pen. Some are drawn to the colours of ink, others are drawn to pen design. For others it's a visceral experience, the feel of the pen on the paper. Or a combination of all of these. The people who use them *want* to use them, and what I'm trying to do is provide a support system." He also supports his family. He works out of his home (he keeps a fountain pen in every room), and his children are part of the business. "My parents had a business in the house and I got used to being around it." Now the family and the extended community of fountain pen users overlap, in a blend of traditional and electronic social networks.

II

At Pendemonium in Fort Madison, Iowa, a store famous to pen devotees, a woman picked up the phone. How retro—not a robo-menu but a live voice. I was talking to Sam Fiorella, the co-founder of the company, and she was working with a Penscapes pen: rubber, hand-painted with a Route 66 theme. She collects interesting pens but is very matter-of-fact about ink. She wasn't sure people were actually buying more ink, just that with the internet we were more aware of it—all those people on Fountain Pen Network writing comments and reviews. There are more colours now, but "black and blue are still the biggest sellers." As for the controversies over ink, she said, "Look, I'm very down to earth about this. There have always been some inks that don't work with some pens. That's just the way it is. If an ink doesn't work in your favourite pen then try it in another one. And if you can't use your favourite shade of blue it's not the end of the world."

I felt a pang. Okay, it's not the end of the world, but the fact that I couldn't use my favourite shade of Private Reserve blue in my Parker 51 pained me. Of course, I could hardly use anything in my Parker 51 and had pretty much given up on it as a working pen because the nib was so fine. Sam said, "Those old pens, most of them have much finer nibs than we're used to now. That's because ink used to be very expensive, and fine points put down less ink." I'd never thought of that. There was another factor: "You get broader nibs, and italic nibs in Britain and Europe," she said. "That has to do with the style of penmanship. Now most pens come with a medium nib."

I started writing with an Osmiroid because my thesis supervisor wrote with one. She favoured a broad italic nib and black ink. When a friend saw her comments on my first chapter he said, "Wow! She must be in love with you." I always liked italic nibs because they were like Botox for my scraggly handwriting, plumping it up a little, giving it shape and shading. Most people start out with a middle-of-the-road medium, but now there's an increased interest in custom grinding, a service Pendemonium offers, so that you can have that medium nib shaped exactly as you want it. I told her I'd send in my Caran d'Ache, which felt wonderful in the hand but produced writing without flair, for a nib grind. As we spoke I was looking at the photograph of her storefront on my laptop, struck by the convergence of technologies old and new: I'd found her on the internet, contacted her by telephone, and was talking to her about pens and ink.

Still, there's no substitute for being there. Although the net brings people together and facilitates one-click impulse buys, there's nothing like a pen store, each one as distinct in character as independent bookstores used to be, and staffed by people who are crazy about their products. Their enthusiasm can be dangerously infectious. In Toronto at Laywine's they said, "Try this Italian pen, a Stipula, from Milan. It has an amazing titanium nib…." Three passes with it and I was making small moaning sounds. At Sleuth and Statesman, three kilometres due south of Laywine's in the financial district, they tell me that young people are discovering "the feel of writing, that you can actually write with emotion." Four kilometres west, in a funkier area on Dundas Street, Liz and Jon had opened WonderPens, geared more toward the introductory market. Like Pendemonium they combined a bricks-and-mortar store with a well-organized web service, with quick shipping at a flat rate across Canada. It's a business that's taking in new blood and that isn't confined to a single demographic.

When my batch of samples arrived from Goulet, I lined up the Private Reserve blues. I liked the saturation of the colours, the flow, the way they worked on the cheap paper of my coil notebooks. I had Electric DC Blue, DC Supershow Blue, American Blue, and Midnight Blues—my ideal of a blue-black. But Electric DC and DC Supershow seemed misnamed (though insiders would recognize that they were named for the world's biggest pen show). The words imply something flashy, something that leaps off the page, but Electric isn't electric at all, it's stately, and I remember why I used it as my daily ink for months when I first bought a bottle. The blue comes through, but it's still subdued. Like a good typeface, it has character but doesn't call attention to itself. I feel it expresses my personality, not someone else's—suitable for a warm note to the dean or a senior editor. If things weren't going well with the senior editor, if I wanted to make my message a little more formal, even stern, I'd turn to Midnight Blues. If I was excited about some new find and wanted to tell my colleagues I'd use DC Supershow—it's a happy blue, but not too casual: you're still at work. With American Blue you're at the lake, on vacation, writing to friends: but still a writer. I also had Tsuki-yo, from Pilot's Isroshizuku series, which puts you at the ocean, watching the green emerge from the depths, inviting your reader to contemplate the surge. Careful, or it will lead you to write pretentious haiku.

But then you wake up the next morning and the inks don't seem to have these associations at all. Electric DC is too bright; Midnight Blues not stern but soft and contemplative. Just what you want. Ink suits the mood as well as the message. I of course told Hsing none of this, but I didn't need to. I'd joined the Fountain Pen Network—one of the 78,000 subscribers (the population of Peterborough, Ontario, or Napa, California), one of the 100 to 120 new subscribers each day. Since coming online in 2004 they had accumulated over 2,500,000 posts in the forums. This was more than discovering a community; this was like coming out of a dark quiet room into a packed discotheque. The volume of information was overwhelming—I understood now why people went to Brian Goulet's videos and Richard Binder's pithy website to make sense of it all. Yet if I ever needed to talk about shades of blue, I'd found the place.

The internet, credited with killing ink, has in fact revitalized it, the new social media enabling the practice of an older social medium. Of course it's not the same. In 1899 Thorstein Veblen observed that candlelight is now "less distressing to well-bred eyes" than it was thirty years before, when candles were the principal means of domestic lighting. With the bright flare of gas lamps and electric lights, candles faded in the utilitarian sphere but glowed as a mark of leisure. They're still the emblem of small dinner parties and of romance; they imply thoughtfulness and intimacy. So too with the pen. The inkwell in the post office has been replaced by a greasy ballpoint on a chain, leaving ink as the mark of the personal. Now in a hue that is distinctly yours.

III

Designing an ink is always a balancing act: finding the formula that will allow it to flow well but dry fast enough, that will be saturated enough with dye to produce a deep hue but not clog the pen (though some argue that dye saturation has nothing to do with clogging). Detractors say Private Reserve's Tanzanite purple "stains badly and flows far too readily"—it's been dubbed "the Ex-Lax of inks," a witticism endlessly repeated, despite all the posts testifying that it doesn't flow any faster than other inks (Hsing loves it and uses it in a purple Delta pen from Italy). The blogosphere of ink can be harsh.

Yet nothing compares to the firestorm of debate generated by Noodler's Baystate Blue: it stained converters, clogged nibs, and melted the plastic feed of some Lamy pens. Nathan Tardif, inventor, manager, and proprietor of Noodler's Ink, fired back in a defiant twenty-nine-minute video in which, along with an ink demonstration, he states that Baystate Blue isn't just an ink, it's an expression of philosophical difference between him and the big manufacturers. He will not be herded into cancelling this ink by "a crowd of so-called Holier-Than-Thous." He brandishes Milton and Rose Friedman's book *Free to Choose* and declares, "It is this book that will cause this ink to be made until I am a dead man, regardless of what anyone says about it online." Debate became so rancorous that the Fountain Pen Network posted a notice:

ALL INK BRAND BATTLES (pro or anti stances) are
banned. Anyone participating will receive a warning and all
related posts will be pulled.... This ban is extended to the
specific ink, Noodler's BAY STATE BLUE.

Not so different from the ink wars of the Ming dynasty. The world
of ink has always been small, passionate, and contentious.

I first heard of Noodler's through *The Economist*, which ran
an article on a "laser-proof" ink. Nathan Tardif had advertised
an award for anyone who could erase his inks without harming
the paper underneath (the test was to foil would-be forgers).
Nicholas Masluk, a grad student in physics at Yale, took up the
challenge and with a laser managed to obliterate one of the inks—
so Tardif developed a new one, guaranteed laser-proof, that he
called Bad-Belted Kingfisher. This was perverse and wonderful.
In the digital age when no one wrote cheques anyway, why devote
your energies to developing a laser-impervious ink? And why call
it Bad-Belted Kingfisher instead of Laser-Proof Blue? Who was
this guy?

A *noodle* in the eighteenth century was a fool, a silly person,
though *noodle* was also your head, and you could be exhorted to
"use your noodle." By the mid-twentieth century, *noodling* had
associations that ranged from "unproductive" to "undirected" to
"improvisational," from slacking off to free-form jazz. For most of
us it signifies aimless writing, or diddling around on the guitar, but
somewhere along the way in the southern United States *noodling*
came to designate the activity of fishing for catfish with your bare
hands. All of those associations are present in Noodler's inks, ink for
noodlers—though if you want to be strict about the apostrophe, it's
the ink of one noodler, the ink made by the arch-Noodler, Nathan
Tardif.

"There's nothing snobbish in noodling," says Tardif in one of his online interviews. "You're down there in the mud, you don't need a yacht, you don't need sophisticated echo location, you can even noodle with no legs. It's the earliest form of fishing in North America." He cackles. It is, I would come to see, a characteristic Tardif statement, combining anti-elitism, a sense of history, and a dash of the grotesque.

Every Noodler's ink label proclaims *Made in USA*; Tardif has created a pen that is American made and sells for twenty dollars. He's fiercely patriotic. He's a man with a mission, and that includes giving value for money, like it or not. He fills his bottles right to the top, even though customers have begged him not to because the air bubble bursts and you may get a fine mist of ink on your face even before you put it in your pen. Retailers' websites now contain warnings to stand well back when opening, though buyers would gladly pay the same for half a millimetre less ink. Bottles filled to the meniscus are part of the Noodler's philosophy of thrift.

However, Tardif's real concern is permanence: "Eternal" inks are designed to resist "the effects of time—moisture, humidity, UV light, acids, water exposure, and many common detergents such as dish soaps...." His "Bulletproof" inks take that resistance further: they're impervious to "all the known tools of a forger, UV light, UV light wands, bleaches, alcohols, solvents, petrochemicals, oven cleaners and carpet cleaners ..." This sounded obsessive. And he has no use for ballpoints: "Fountain pen ink should outlast the faded ballpen ink we now see before us in archives from the 1960s and 1970s. Fountain pen ink should still be on the page when your grandchildren read what you wrote." His ideal is to reach out "as far into immortality as the written word permits us to reach— the only way to speak to future generations remains the written word durably made upon the page." He even makes a "Polar" ink, good to minus thirty degrees, perhaps a nod to his Canadian background. (I leave a bottle in an unheated mountain cabin, and it always works.) Noodler's inks have high dye content and have been criticized for staining pens and clogging nibs, but Tardif insists, "Weaker inks make fountain pens useless against forgers ... let alone simple raindrops when mailing a letter."

The words conjure up a time and a practice long past. Never mind forgery, when was the last time you mailed a letter in the rain? Intrigued, I bought some Baystate Blue, inked up an old Lamy (pen aficionados refer to their pens as being "inked" rather than "full of ink"), and phoned Nathan Tardif. It turns out that rain figures prominently in the Origin Story. "I had a friend who said, 'How many bottles of ink do I have to buy for you to produce a permanent ink for my crossword puzzle? So when I'm on the subway and I get off at an open-air stop and it's raining and I have to use that paper as an umbrella, that crossword isn't going to be wiped out.'

"I said, 'Fifteen to thirty-six bottles,' and he said, 'Done deal,' and that was the first customer. That commitment, small as it seems now, was enough to make me take the risk, because I figured if he would do it other people would." The opposite of a dot-com startup. He's never looked for millions of users.

Asked about the laser experiment, Tardif says, "I found [Nicholas Masluk, the Yale grad student] very impressive. Microscopically he had to knock and burn each particle of ink off on the paper fibre." He had to be careful because "the paper fibre would burn a tiny amount and leave a yellowing so even that would give it away to a teller with a magnifying glass. It was fascinating that he'd gotten to this microscopic level." I was fascinated that Tardif was fascinated—these two seemed made for each other. In the Middle Ages they would have been alchemists. The ink world attracts obsessives.

He's a loner, and you suspect sometimes lonely, but he's never going to work by committee, or with a design team, tossing a baseball and ideas back and forth across a table. Ink dealer Brian Goulet calls him "the Willy Wonka" of the fountain pen world.

Tardif says, "I've been around in the hobby a lot but never felt I was one of them, nor ever became one of them. Some claim that I'm Satan incarnate, that I'm the worst thing that ever happened to the fountain pen. But if I'd followed the conventional rules I would never have experimented with the new ink concepts that were being developed in the 90s."

I point out that he's become a global phenomenon, and this triggers a rant. He'd just received a letter from some agency in the European Union: "They want me to disclose my formulas in Europe and they want me to divulge how I make it. I would say 'Tell Coca-Cola to divulge its formula first!' I can't stand this idea that the government can force you to divulge trade secrets. It's insanity."

He segues to Massachusetts labour laws. "I could hire thirty people on the spot, but the government has made the economics impossible! The health care costs are so high I have to avoid labour like the plague. I have a joke I tell people, 'The state of Massachusetts has made it illegal to hire people!' They want the highest minimum wage in the country." Which leads the conversation to education.

"I encounter college graduates working in coffee shops who can't give me change for a two-dollar bill because there isn't a pictogram for it on the cash register. They're stumped, they're flummoxed, the computer doesn't compute and the whole thing collapses. That's our education system! As a test I ask them, 'What party affiliation was Abraham Lincoln?' Eight times I did this. Eight times they told me 'Democrat'! I blame the teachers' union!"

His standards are those of another era. "If you want to wander away from the 30s and 40s and the standard they had—which was a nice line, well defined, high contrast, you can read it without your glasses—this will lead us back to dish soap–weak ink of the 1970s." He started designing his own pens with ebonite feeds that, like vintage pens, can handle any ink. He wanted fountain pen ink to be practical again, "capable of doodling ideas and concepts in as many places and forms as is possible for a fountain pen ink … an ink to write ideas on the airport meal sheet, back of the newspaper, or on an insert torn from a magazine." He sounds like Laurence Bich on the utility of the ballpoint pen, except that he's reversing the historical flow.

"People tell me, 'Your ink takes an eternity to dry.' And it's a fourteen-second dry time. Here you're waiting for the ink to react with the paper so that your words will last an eternity, and you won't wait fourteen seconds!" But fourteen seconds *is* an eternity in the digital age. One pen shop manager observed, "Nathan seems to be a kind of throwback to a time when you'd be driving down the road and you'd pick up somebody hitchhiking and you'd get to talking and you'd decide to have that person work for you. It was a time when a handshake meant something, and you weren't bogged down in all the paperwork we have today. It's a kind of nostalgia for an era that doesn't exist anymore."

His criterion is value for money: "Why would you spend twenty dollars for an ounce of J. Herbin from Paris when you could get twice that for the same money?" But this dodges the fact that Noodler's ink is itself a luxury good, especially if you're going to indulge in different colours—which the range and the labels plainly tempt you to do. Most fountain pen geeks already have more ink on their shelf than they can use in a lifetime.

What attracts people to Noodler's isn't the laser resistance or the topped-up bottle, it's the labels. Allusive, quirky, provocative, each one generates a story. The historical themes range from the First Chinese Emperor to Tiananmen Square, from Lawrence of Arabia to Abraham Lincoln, from Julius Caesar to Oppenheimer and the atomic bomb to the bear that inspired Winnie the Pooh. Fans on the web write that they'd love to own them all, even if they never opened the bottles. Tardif could stamp *Not For Use* on the bottle as Fang did with his most exclusive inksticks, and they'd still sell. I surveyed a dozen labels, and taken together, they express an interest in history, a fascination with chemistry and colour, a deep political commitment, and an equally strong spirit of whimsy. Most are designed to rile at least somebody: aristocrats, Democrats, prudes, historians, Chinese Communists— even Canadians.

Le Couleur Royale features Louis XIV on the guillotine, his head and entrails being shown to the crowd. "Three retailers sent it back, saying 'This is outrageous!'" Tardif cackles. "I'm cynical about aristocrats: my ancestors used to be dunked on Boston Common often enough; my Canadian side was French Canadian; my British side was revolutionary—one of them was part owner of the ships at the Tea Party. Maybe it's a genetic thing, but I innately despise the class structure, I believe in a meritocracy…. I was poking fun at Louis XIV, a complete incompetent." (J. Herbin inks, on the other hand, proudly announce that their founder made ink for Louis XIV, and later for Victor Hugo.) Tardif laughs again, and says, "Making a controversial statement is profitable to the soul." He seems to go on a lot about value for money, though in strict business terms it would make more sense to keep bottle and label the same and just put a little identifying sticker on the cap, as Diamine does. Yet in artistic terms the labels are essential. Like Cheng and Fang, he wants to do more than just make good ink.

Sometimes Tardif allows his inner fourteen-year-old to take over. He tells me with a laugh how some dealers took offence at Ottoman Rose, which displays a harem with bare-breasted women, and Mata Hari's Cordial, condemned as "sophomoric" because "I made Mata Hari, the famous spy, into a mermaid, and though she was famously flat-chested I told the artist to give her an incredible endowment … with outstanding nipples. The British objected— 'Mata Hari had small breasts, this isn't right!'—and I said, 'Well she wasn't a mermaid either!'" Tardif cackles again until he wheezes, still enjoying the joke.

Yet some of his labels reflect weeks of research. For the bright-red First Emperor Ink he corresponded with museum curators, trying to discover the colour of the ink used in the Imperial City. He

Nathan Tardif, creator of Noodler's Ink, acknowledges that
Mata Hari, the famous spy, was neither full-breasted nor a mermaid.
© *Nathan Tardif, Noodler's Ink.*

talked to them about inksticks, and then about inkstones. "They had more data on the inkstones than the inksticks, and I figured the scrapings on the stones would be more likely to be authentic." Then, through the National Museum in Taipei, he discovered fragments of the pigment used on the famous terra-cotta warriors of the so-called First Emperor, Qin Shi Huang. Tardif was fascinated by the chemistry. "It fluoresced. The diggers saw the colours, but as soon as it came out of the ground it oxidized and faded away. So they found a way to preserve it, and they put all the data online and I got the colour." So he created an ink that goes on blood red, like the colour of the Emperor's chop marks, and dries the colour of red used in the tomb. Also, it will fluoresce under UV light.

Typically for Tardif's inks, the First Emperor Ink label doesn't tell you the colour, but this one doesn't even tell you the name. On the left side of the bottle are three characters in red: if you read Chinese, or if you've seen the video, you'll know that these represent the name of the Emperor, Qin Shi Huang. If not, you're out of luck. Only by turning the bottle this way and that do you come to see that the image is a boat, out on the ocean. The sea is blue, the sails of the ship (not rugs, you now realize) are patterned red and orange, with blue highlights; the hull of the ship is black. On the left side, above the prow of the ship, a green flag waves in the wind. And suspended in the sky is a green and red oblong with Chinese characters on it: a Chinese inkstick. I didn't recognize it at first, and it was an odd sensation when I did. I'd gone all the way to China to search for artisans devoted to ink, and had come back to find one here.

This is the ship the emperor sent out to search for the secret of eternal life. Tardif told me, "I liked the story of the young people in the boat going out to seek immortality. From a marketing perspective I wanted to get something that would enable the mind to wander with the label as much as possible, as well as to get the ink out the door." I agreed that it was a beautiful story— five hundred boys and five hundred girls set out, found these islands and, legend has it, founded Japan. But I'd read about Qin Shi Huang. I told Tardif that his emperor was the one who wiped out all the archives because he wanted world history to begin with him. He feared ink: he didn't want the records of the past used to criticize the present. Worse, he killed hundreds of scholars by burying them alive (in part because a couple of alchemists had deceived him about immortality). Further, these thousand boys and girls on the ship, they weren't going to be the ones to achieve immortality—they were going to be sacrificed to the gods to guarantee that the emperor lived forever. I wondered if Tardif knew that.

"Right!" he cackled. "Everything about him was horribly brutal. He's not someone you'd want to invite to dinner." He paused. "The ink is eerie, because it moves like blood in the bottle, and he's an eerie character." Tardif admits he'd had reservations: "I hope that one isn't too creepy. When I first bottled it I got a chill down my spine." When you open the bottle and look at the froth on the inside of the lid, it's the colour of bright arterial blood bubbling up out of a fresh wound.

Tardif includes sly references to his Canadian heritage. On the laser-proof Bad-Belted Kingfisher, the bird says *"Je parle canadien,"* and Tardif tells me about his Blue Upon the Plains of Abraham for the Canadian market, for the Sleuth and Statesman store in Toronto. "It's got that portrait of Montcalm, the French flag, the British flag, and," he laughs, "right in the middle of the battlefield there's a Canadian flag." One website solemnly takes up the histor- ical inaccuracies of the flags. "It was a bi-gravity ink that separated into three colours—blue at the top, the colour of the fleur-de-lys, then a mixture that was burgundy, then British red. If you let it settle, by the end of the week you could fill from any three of the layers. Being French, I put the blue at the top." He laughs. I tell him that this is the most explicit instance I've encountered of ink itself being a political statement, and ask where I can find a bottle. He doesn't know, as he retired the ink. "I discarded the formula because it really didn't work, and anyway the market isn't ready for bi-gravity inks."

The labels are brilliant promotion. Ink enthusiasts discuss them on the net and customers are drawn to them in stores. Italian writer Umberto Eco says that a good book title should slightly mystify readers, the better to draw them in. Tardif seems to have absorbed the principle, forcing you to discover the colour and ponder its connection to the name. Like the designs on Ming dynasty inksticks, the Noodler's labels take you far afield, to dispatches of the military, the edicts of emperors, the antics of a bear, the search for a white whale. They reveal a jackdaw mind, attracted by bright bits of information.

Tardif is obsessed with "cheque washers." He explains how they can wax-coat the signature and wash off the rest from a cheque without affecting the security lines. This is steam-punk paranoia; these days people are more concerned about computer hacking than cheque washing, but with the issue of permanence he's on to something. He says, "How much confidence does the average person have that their writing is going to be read by their grandchildren if it's written on an iPhone? It's a catastrophic loss when a hard drive goes. At the end of the day what you have is what you printed, what you wrote." Computer users have all faced the blue screen of death, and remain haunted that everything they've done can vanish in an instant. It's not monetary loss so much as identity loss that we fear. Part of the attraction of ink is that it's a way to foil not cheque forgers or government bureaucrats but the Grim Reaper himself; it's a way to cheat death.

Tardif is the opposite of Marcel Bich. Profit isn't his motive, nor is world domination. He reminded me of Lazlo Bíró, another outsider, more inventor than businessman (though Tardif's college training was all business), who regarded pens not as jewellery but as a force for democracy and the obliteration of class barriers. For Tardif, mass culture, especially throwaway culture, is anathema. Cartridges are evil. "If they get out into the sea they float around and look like something edible; fish and sea turtles eat them. They kill things. Noodler's will never produce a cartridge." For him the pen represents freedom from the corporate matrix: "The fountain pen has no monthly contract." And the pen represents simplicity: "You don't have to boot up, turn on, plug in a code—you just take the top off and write!" Until we had cell phones and computers we didn't see this as a virtue; now it seems like a brilliant innovation. Yet for all Tardif's attachment to the past, the phenomenon of a global one-man ink company would only be possible in the digital age. And his success is proof, if any more were needed, of a global hunger for ink.

IV

The number of calamophiles continues to grow. *Calamophile*—I learned the term from a blogspot—means "pen lover," and derives from *calamus*, Latin for "reed writer." These pen lovers get together not just online or at pen shows (which may draw as many as two thousand people) but also in smaller groups, in local pen stores or in each other's living rooms. There are long-standing clubs, such as the British Writing Equipment Society, inaugurated in 1980, or the Pen Collectors of America, a national organization that grew out of the Southern California Pen Collectors Club in the late 1980s. Since 2000, however, many new clubs have formed, and although Canada doesn't have a national group, pen clubs have started up in Vancouver, Calgary, Ottawa, and London, Ontario.

For one-stop text on pens, one of the best sites to consult is RichardsPens. Richard Binder is a man who loves pens. He writes well; on the website and in his Glossopedia (a glossary and encyclopedia) he lays out the intricacies of the fountain pen with a clarity that would make Cennini or Moxon proud. When I spoke with Richard on the phone I mentioned that I'd seen an ink described as "water-colour" and had come to realize that I liked inks that were instead thick, dark, opaque, almost like printers' ink. Now I was beginning to wonder what that said about me. Did it make me less sensitive? Was it the equivalent of preferring a syrupy shiraz to a delicate pinot noir? Were people snobbish about inks? He dodged the metaphor and my ink anxiety. He did acknowledge that "there is snobbery associated with brands of ink," but as for the intensity, "Those who prefer inks that will give you shading (the variation in colour) are those who are more interested in the artistic effect. The ones who prefer the solid heavy colour are the ones who are after the drama."

Actually I just wanted an ink that lays down a good line and is easy to read. I said that after talking to pen-store owners for a while they'll sometimes admit that their daily ink is something very basic, such as Waterman's Blue-Black. Binder likes "low-maintenance ink," specifically Waterman's Mysterious Blue (which, he explained in a verbal sidebar, until about two years ago was Waterman's Blue-Black, but there hadn't been any black in it for years because they changed the formula a couple of decades ago). "Mysterious Blue is my favourite ink, not because I love the colour to death, but because it flows well, it lubricates well, it doesn't stain and it doesn't clog. What more can you ask of an ink?" He'd never sacrifice utility for colour. In fact, he said, "I can't honestly recommend you use any Noodler's or Private Reserve inks." His website contains a boldface warning about these two "boutique inks" that create problems ranging from "flow issues and clogging, mold, staining" to "actual destruction of pens."

I told him that I needed a fountain pen just for readability; with a ballpoint my handwriting is a manic scrawl—five minutes later I can't decipher what I've written. He told me why this is so. "The National Education Association has done a study that explores, among other things, how ballpoint pens are a major contributor to bad handwriting." The difference between fountain pens and ballpoints is in the way the pen is handled. "With a fountain pen you hold it in a normal, relaxed manner. If you learned properly you'll be holding it gently, with just enough pressure to keep it from slipping out of your hand…. With a ballpoint you can't hold it at that relaxed angle, because if you do you're going to be writing on the side of the refill where it's crimped to hold the ball in. So you raise the angle of incidence. That's not a good thing because it changes the shape of your hand, and it's not a natural position. Then in order to make the pen write you have to push down on it. Which is bad because to push down on it you have to squeeze it harder. Which tightens up all the muscles in your hand. The result is that you're holding the pen in an unnatural way, tightly, uncomfortably, and just to make it make marks. None of these is a flaw of the fountain pen."

As for the demand, "I don't think the market is going to explode. I suspect there's gentle expansion going on. The computer and the devices that go with it, the iPhones and iPads, have created a social disconnect—people don't have the human contact that it turns out we all need. Some of the young people are recognizing this, and they're responding by using pens and paper because they're closer to other people that way." Again there was that notion of the material object, touched by the body, as a proxy for physical contact.

Some long-time pen shops are closing, but others are opening with the entrepreneurial fervour and flair usually associated with new restaurants; it's a business fuelled by passion, not the bottom line. Some pens signify that they're for CEOs to sign contracts with while the worker bees scribble with their Bics—perhaps a Montblanc Meisterstück still says "I'm rich, I have money and leisure time to burn"—but a fountain pen relates to time in another way. Filling the pen creates a still point. As the pen breathes in its ink you may hold your own breath, or consciously adjust it to the motion, but for those few seconds you're absorbed in watching the column rise, careful not to spill. For a Westerner, in our speeded-up culture, it's the equivalent of grinding an inkstick one hundred times. It slows you, readies you. The grandmother in Amy Tan's book would grudgingly approve.

In 1492, at a time when the printing press was taking over the market, a monk named Johannes Trithemius pushed back: "though many books are now printed," he wrote, "you will always find some that need to be scribed." His *In Praise of Scribes* was spectacularly wrong. The world was changing under his feet— the Moors were being pushed out of Spain, Columbus was about to stumble upon North America, and the printing process that Gutenberg had unleashed some forty years before (only 130 kilometres from Trithemius's hometown) had spread across Europe. The time of the scribe was dead and gone. Was I, in my enthusiasm for ink, like Trithemius, unwilling to recognize the end of an era? Children now often learn to use a keyboard before they use a pen, and many schools no longer teach cursive writing. We text, we emoticon.

Yet in some MBA programs they instruct students in the art of the handwritten note. An email from your boss saying "Good job, Jill" or "Nice work, Fritz" might be nice, but on a slice of half-size stationery, handwritten, it becomes an artifact worth saving. Five years ago half the students in my class had laptops; I was certain that they'd all be using computers soon. Pens were for opening bags of chips. Today, however, in any class of forty, only two will have laptops. Students are using notebooks and ballpoints. Some are even turning up with fountain pens, not expensive grad gifts but solid working pens. They tell me, "Yeah, I get more down with a computer, but I don't remember anything. When I write with a pen it seems to stick." One day while teaching in my university's term-abroad program in Cortona, Italy, I found my students in a café addressing piles of postcards, with cell phones and laptops open beside them. Why not text or email? "It's not the same," said Celestia. "A postcard is a material memory," said Paige. If you want to exchange information, you do it digitally. If you want to preserve information, you put it in ink.

I once brought my grandfather's Smith Corona typewriter to class, confident that I'd amaze and confound the students by making them try to type. Joyce Tam, a young woman in the humanities computing program, fired off a sentence, making the keys clack like a machine gun. I was the one amazed. She shrugged. "After spending all day in the lab designing websites, I want something concrete. I have an Underwood 5 at home." At the local espresso bar I ran into Stan Ruecker, the head of the computing program, and told him the story. He was surprised at my surprise. "Of course," he said. "The virtual breeds a hunger for the actual."

Gum arabic is crystalline and shatters easily,
but gallnuts require a sharp whack. Photo courtesy of the author.

Making It

I

W.H. Auden paraphrased French poet Paul Valéry: "Poems are never finished, only abandoned." What Valéry actually said was less elegant, more accurate: "A work is never completed except by some accident such as weariness, satisfaction, the need to deliver, or death…." (*Un travail n'est jamais achevé, sauf par accident, comme la fatigue, la satisfaction, la nécessité d'offrir ou de mort.*) I think he knew that satisfaction was an illusion, but wanted to posit the possibility for some writer somewhere; death we have no control over; even weariness is inevitable but not terminal: the writer, sated, sickened, still prods listlessly at the project, reluctant to relinquish. But the need to deliver, yes. I would have to abandon much.

The trail of ink left tantalizing leads and intriguing characters unpursued—like Mr. Francis Wells, the English cardiac surgeon (British surgeons proudly refer to themselves as "Mr.," not "Dr.") who sketches in the operating room with forceps, using blood. As he told *The Guardian*, "I must say it's a wonderful drawing medium. It flows beautifully." His work was included in a summer exhibition of the Royal Academy, reproduced "disappointingly" in ink and ballpoint. ("Blood can be dangerous stuff," he said.) Then there's Mark Gruenwald, the graphic artist at Marvel Comics who requested that on his death he be cremated and his ashes mixed with ink to print the paperback version of his comic *Squadron Supreme*.

There's also a whole field of invisible inks and ingenious frauds. The "sharpers of the racecourse" devised betting tickets written in two kinds of ink, one that faded away while the other gradually appeared—if your horse won, you'd go to claim your money only to find that the ticket didn't say what you thought it had. (If you want to try this, make the first with a weak solution of starch and a tincture of iodine, and the second with an ammoniacal solution of silver nitrate, which gradually darkens when exposed to light.) Even the U.S. Secret Service has a file of over eight thousand inks (clearly I needed to make a field trip to Langley). Gallnuts kept popping up: Hitler's diaries, bought by the magazine *Stern* in 1983 for four million dollars, were proven false on the strength of the ink. Instead of gallnut—the standard ink for official documents in Germany during the Second World War—the diaries had been forged in a contemporary Pelikan fountain pen ink. On the other hand, the "Judas Gospel," a third-century gospel discovered in the 1970s that portrays Judas as co-conspirator rather than betrayer (in this version Jesus requests that Judas turn him over to the authorities for execution), was proved period-authentic in 2013 based on the composition of the iron-gall ink.

Forgery of a different sort lies at the heart of the famous "Indian Yellow" case. The vibrant paint, so prized by nineteenth-century English painters, was made from the urine of oxen that had been force-fed mango leaves. Oxen don't normally eat mangos, and they withered under the treatment—which led to protests in the 1880s and then to the passing of laws in the state of Bengal forbidding the production of Indian Yellow. I'd encountered the story several times. Then I came across Victoria Finlay's *Color: A Natural History of the Palette*, in which she describes how she searched Calcutta's National Library and London's India Office Library and found nothing. So she set off for Monghyr, India, where she talked to locals using a sketch of a cow, a mango leaf, and a urine bucket. No one knew anything about the alleged process. Finlay concludes that the whole thing was a hoax, the Indians taking the piss out of the British, maybe as part of the period's rising national consciousness and its new movement to preserve the sacred cow.

Or maybe it was a metaphor for the indigo trade. There's a story about an Egyptian who caught a stork, dyed it blue with indigo, and offered it to the French consul as a rare bird, but the tale of colonialism and indigo wasn't often funny. So-called "Indian" ink wasn't Indian: any ship that had travelled from the East around the Cape of Good Hope was listed as coming from the "East Indies," a vague term that could mean anything from the subcontinent of India to China. So carbon black ink, although manufactured in China, came to be known as Indian ink. But there was a component of blue ink that did come from the subcontinent: indigo. The name derives from the Greek *indicon*—from India.

Thousands of acres of the plants were grown in West Bengal, as well as in other areas of the world, including Louisiana, where until 1763 indigo was the primary export. The indigo trade boomed as manufacturers used it not just for ink but for textiles—military uniforms in Europe and Levi Strauss's blue jeans in the California Gold Rush—but through the 1800s European scientists were competing as fiercely, and secretively, as software developers to create a synthetic indigo. The Germans finally succeeded, and in 1878 the bottom dropped out of the indigo market.

Like cotton in the American south, indigo—known as "the Blue Devil"—became synonymous with oppression. British planters fined, beat, or burned down the homes of farmers who refused to grow indigo, and pushed through an order forbidding rice to be grown on lands that had ever grown indigo, even though indigo prices were falling and the peasants were starving to death. When the peasants rioted, the authorities brought in a young lawyer to mediate: Mahatma Gandhi. The techniques of civil disobedience used in the indigo uprisings provided the template for the movement that would ultimately achieve Indian independence. Later the first political protest play written in Bengali, *Neel Darpan* (*The Mirror of Indigo*), dramatized the uprising of peasants, an instance of the ink compound feeding directly into politics and literature.

Ibn Badis, the prince who in 1025 wrote *The Staff of the Scribes*, used indigo in several of his ink recipes, usually in combination with some other substance to produce blue-greens. Indigo was used for staining leather, parchment, and paper. Cennino Cennini has a number of recipes that use indigo: like lapis lazuli, it was imported from the Middle East and thus costly; it would have impressed wealthy patrons. Cennini sometimes specifies "Baghdad indigo," though as with so many products, Baghdad was just the marketing and distribution centre. In early sixteenth-century Venice, blue paper, *carta azzura* or *carta turchina*, became the fashion. This influenced the deluxe printing of Hebrew religious books—but the most fabulous example is the Blue Qur'an.

Crafted in the ninth century (a century or more before ibn Badis), the Blue Qur'an's pages are a goat-skin parchment dyed dark blue with indigo. The lettering is in gold ink outlined with black gallnut ink, and the chapter headings were in silver, now oxidized to black. With just fifteen lines to the page, it was a massive volume—650 sheets—and was designed for display rather than reading. It would have been enormously expensive, and remains so today. Individual sheets surface occasionally in the art world; in 2012, one sold at Christie's in London for $390,000. The last time it was recorded as being intact was in 1293. Today it's housed in Tunisia, where I made a fruitless trip to try to view it, arriving, all unaware, on the anniversary of the date when a fruit seller named Mohamed Bouazizi had set fire to himself, igniting the revolution in Tunisia, catalyst for the Arab Spring. The museum was open, but surrounded by razor wire, and the staff were unforthcoming.

II

I told Hsing, "I need to return to Tunisia, I need to go to West Bengal, I need to get a tattoo!"

In Shanghai I'd wanted to get a tattoo of a Chinese character, done in the calligraphy style of Dong Qichang. The custom used to be that you'd get a tattoo far from home, a memento of some exotic locale. The Shanghai tattoo parlours looked as sterile as operating rooms, but Hsing, who despised tattoos, assured me there'd be cost-cutting measures that would compromise the process. What finally dissuaded me was the smog, a toxic porridge that left me coughing for weeks. It was bad enough breathing it; I didn't want to puncture myself and let it under my skin. Still, the whole notion of tattoos had gotten under my skin.

As more and more of my students were coming to class with electronic devices, more and more were displaying tattoos. I couldn't prove a link, but this seemed the ultimate corporeal counterpoint to the digital: where a power surge could wipe out years' worth of notes, only surgery could eradicate the skin's inscription. But what were the protocols in reading these texts? One student who sat in the front row had a bird on her chest, the wingtips just below the collarbone, and the eye was drawn there. I knew not to say "Nice bird," but do you say "Nice ink"? (*ink* being the generic term for tattoos, I learned), or do you comment at all? The bird was there all one semester, highlighted by scoop-necked tops, but the next it had disappeared behind high-necked blouses. "There are basically two kinds of tattoos," my students instructed me. "Those you can see and those you can't." Yet the tattooed body can be like tidal flats, sometimes covered, sometimes not. To view any tattoo is an intimate act, and facial tattoos are the most provocative of all, literally in your face if you're confronting the wearer. Then it's the intimacy of the warrior.

The history of tattooing goes back centuries: the Romans tattooed their slaves; the Celts tattooed their warriors (the Romans, daunted by their facial and body tattoos, called them "Picts"—from *pingere*, to paint). King Harold II, killed at the Battle of Hastings in 1066, apparently had *Edith* tattooed over his heart. The basic technique is the same today as it was for King Harold: pigments are forced under the skin with a pointed object. Before Samuel O'Reilly's 1891 electric tattooing machine (based on Edison's electric pen, a device that perforated paper for making stencils), all tattooing was done by hand, using small holders, like a pen holder, with a number of needles set in. Modern prison tattoos, using ballpoint ink, preserve the tradition.

In 1774 Captain Cook returned to England from Polynesia with the tattooed Ra'iatean prince, Omai, who was introduced to Samuel Johnson, King George III, and other aristocrats. Sir Joshua Reynolds painted his portrait, and tattooing became popular among the Mayfair set. In 1862, the Prince of Wales, later King Edward VII, received his first tattoo; Lady Randolph

Churchill (mother of Winston) had a snake tattooed on her wrist. But after the turn of the century, the upper classes began to turn away from tattooing; the practice became associated with sailors, circus performers, and prostitutes. Lady Churchill covered her snake with a wide bracelet. By mid-century, although Dorothy Parker had a star on her elbow and George Orwell blue spots on his knuckles, tattoos had slid further down the social scale, now linked in the public mind with concentration camp inmates and gang bikers. This is what Michael Atkinson, Canadian sociologist, tattoo enthusiast, and author of *Tattooed: The Sociogenesis of a Body Art*, calls "The Rebel Era," 1950–1970, when tattoos marked the fringes of society.

Attitudes shifted in the 1970s (the "New Age Era"): a tattoo was no longer a mark of deviance. Many women got tattoos, seeing the body as "a site of self-determination, liberation," and a new cultural aristocracy—film stars, sports heroes, pop singers—made tattoos fashionable again. We're now into what Atkinson has dubbed the "Supermarket Era," when everything is available and tattoo shops no longer lurk down dark alleys. Cate Blanchett (following Lady Churchill?) has a wrist tattoo, and Angelina Jolie has tattoos in Khmer script, in Latin, and in Arabic, along with images and geographical coordinates for the birthplaces of her children. Writers write on their bodies as well as the page: American author John Irving has a maple leaf tattooed on his shoulder (for his Canadian wife), and wrote *Until I Find You* (2005) about a Canadian actor moving through the tattoo subculture; Canadian writer Alissa York (shortlisted for the 2007 Giller Prize for her novel *Effigy*) has a gecko on her left upper arm; and Corey Redekop got a tattoo of the cover of his first novel. The Writers' Union of Canada doesn't collect such data, but there must be many tattooed writers, each with a story.

At the Banff Centre I met Jen Rae, who was doing a visual arts residency that combined interviews, scholarly research, and her own full-body tattoos for her "Intangible Skin" project. I came across a website for Edmonton-born Krystyne Kolorful, a Guinness Book of Records holder for the Most Tattooed Woman. She has tattoos over ninety-five percent of her body, including thirteen dragons. I learned about Shelley Jackson, who, raised in a family bookstore in Berkeley, California, had published print books, then under the influence of early hypertext theorists George Landow and Robert Coover, had produced *Patchwork Girl*, a reworking of Mary Shelley's *Frankenstein* as an ebook that the reader has to patch together herself. Now she was working on the "Skin Project"—a call for volunteers to each have one word of a 2095-word story tattooed on them. At last count she had 1875 words in circulation.

In *Body Type: Intimate Messages Etched in Flesh*, Ina Saltz notes the emergence of the "tattooed affluent": educated young people, often design-school graduates, who are choosing typographic rather than imagistic tattoos. These are often literary passages, such as the famous ending of Joyce's *Ulysses*: "yes I said yes I will yes"; on his left arm Johnny Depp has *silence, exile, cunning* (the watchwords of Stephen Dedalus in Joyce's *Portrait of the Artist as a Young Man*). In *The Word Made Flesh* (not a Bible study), Eva Talmadge and Justin Taylor document "literary tattoos from bookworms worldwide"; these include passages from Rimbaud, Eliot, Vonnegut, Blake, Melville, and Bolaño, not to mention the Dewey decimal number for poetry, and, on Jennifer C. from Calgary, a line from *Harriet the Spy*.

These texts can sometimes go wrong. While working a charity fundraiser at a casino I saw a very tough-looking guy (shaved head, scarred face) with *Your Next* tattooed on his bulging bicep. "Your next what?" I said to my co-worker. "It should be YOU'RE next. The guy's a walking grammatical error! I'm an English professor. Am I honour-bound to correct him?"

"Shut up," said my friend. "He'd obviously kill you."

Maybe so. There's a story about a San Francisco tattoo artist who got fed up with guys coming in for badass tattoos in Chinese script they couldn't read, so instead of *I will crush you* he etched *I am a wimp*, for *Ninja Warrior* he inked *Mamma's Boy*, and so on. When the tough guys paraded through Chinatown in their muscle shirts they got chuckles instead of respect. Eventually the tattoo artist had so many enraged clients that he had to leave town. This may be an urban myth, but it is a cautionary tale for getting a tattoo in a language you don't understand. The Sanskrit prayer on pop-star Rihanna's hip is misspelled, as is the name on David Beckham's arm. Even the rich and famous need to be copy edited.

Whatever the image, sex is the subtext. In *Polynesian Researches During a Residence of Nearly Eight Years in the Society and Sandwich Islands*, Reverend William Ellis (1794–1872), a missionary in the South Sea Islands, gives an account of tattooing's mythic origins. A daughter, called Hinaereeremonoi, was born to the gods: "As she grew up, in order to preserve her chastity, she was made *pahio*, or kept in a kind of enclosure, and constantly attended by her mother. Intent on her seduction the brothers invented tatauing, and marked each other with the figure called Taomaro." The sister admired the tattoos so much that she broke out of the enclosure, was tattooed herself, and "became also the victim to the designs of her brothers."

One anthropologist argues that while the bleached body was considered sacred, tattooing rendered the body less so, more available for sexual intercourse, even incest. In a famous 1920s Boston rape trial, charges were dropped after it emerged in court that the victim had a butterfly tattooed on her leg, certifying that she was "a person of previous sexual experience" and had thus misled her two rapists. And according to Margot Mifflin in *Bodies of Subversion*, the circus sideshows of tattooed women that began in the 1920s (concurrently with the Miss America beauty pageants) usually included a titillating backstory of kidnap and tattoo rape by "savage" Indians. The sexual aspect of being tattooed, penetrated by ink, is foregrounded when it's accepted by a woman but denied completely when it's done to a man (those two Polynesian brother gods weren't opening up to each other were they?), and the most innocent-seeming iconography can be read sexually: a 1972 *Police Gazette* article argues that the butterfly tattoo means you want to be chased and caught. However, the wheel has turned. Women are creating what Christine Braunberger calls "monster beauty"—a combination of aesthetics and anger that has generated such works as Stephanie Farinelli's tattoo: a necklace of "a stunningly colorful variety of thickly veined penises." (We'll just pause here while the male readers take that in.) Braunberger tells of a lawyer who on her sixtieth birthday got a tattoo and was disappointed when the first question from her (enlightened, professional) female friends was "What did your husband say?" Appalling, true. Yet I'd often heard male acquaintances begin the narrative of their tattoos with "Yeah, my wife wasn't keen, but …" Ownership of bodies was an issue too fraught for me broach, the shifting signs of the tattoo subculture too complex for me to unravel. Books, magazines, international conventions continued to proliferate.

Even the ink itself resisted easy classification. For Health Canada and the FDA it falls under both cosmetics and the colour additives in food, so what's deemed safe to go *on* your skin may not be safe to go *in* your skin. There are new organic inks in which the pigment is suspended in some kind of organic solution, and acrylic inks in which the base is a plastic resin. The acrylics are brighter but that gorgeous red may contain mercury; tattoo inks are proprietorial (as always, the components of ink are a trade secret), and so manufacturers don't have to label the contents. Red and yellow are apparently the most common colours to set off an allergic reaction, while black is the safest. A 1928 report entitled *Tattooing and Face and Body Painting of the Thompson Indians, British Columbia* noted that they used a needle made of bone or cactus spines and a thread blacked with powdered charcoal. Things were simpler then, and the pigments were drawn from natural substances. But I'm always suspicious of the good-old-days argument. Maori tattoos used caterpillar fungus and dog fat; early American tattoos used soot for black, brick dust for red, and possibly urine in the mixing. (I'm all for urine in a Gutenberg Bible, but I wouldn't want it in my arm.) The tattooist would moisten the dry colour by putting the needles into his mouth; syphilis was often spread as a result.

All puncture is a risk. One tattoo artist pointed out that tattooing has been a mainstream practice for thirty years now, with no extreme reactions to the new inks; that in itself is the testing. Others counter that three decades isn't long enough to assess what might happen to the body. Then again, you might get hit by a bus. As with those other acts of penetration, part of the attraction is the danger. Hsing told me that if I had a tattoo I couldn't have an MRI because the machine is a giant circular magnet that induces a magnetic field in the iron oxide of the tattoo ink and can cause electricity to run through it. I imagined the tattoo being ripped out of my flesh and stuck to the top of the MRI tube like a beached starfish as I screamed and bled below. (It turns out that the reaction isn't so dramatic, though it does range from itching to tingling to actual burns.)

In any case, I decided to defer the Blue Qur'an, abandon the case of the mango-fed oxen, and postpone the tattoo.

III

I'd read endlessly about ink, in handbooks, guidebooks, encyclopedias, and the word now leapt off the page at me in novels. I noted that Molly Bloom, the heroine of James Joyce's *Ulysses*, had trouble with her handwriting (she "interrogated constantly at varying intervals as to the correct method of writing the capital initial of the name of a city in Canada, Quebec") and treated her pen badly, abandoning that "implement of calligraphy" in the "encaustic pigment, exposed to the corrosive action of copperas, green vitriol and nutgall"—she used gallnut ink. As Steve Pratt had said, scholarship should be married to craftsmanship. It was time I experienced that corrosive action myself.

I asked Stephanie Watkins, a conservationist at the Harry Ransom Center in Austin, how to make gallnut ink. "Go to the Ink Corrosion website," she said. "And don't worry about precise measurements. This is cottage science." True. The "careful housewife," writes Mitchell, "would rank it among her duties to make ink, just as she made cordials, and compounded medicines of marvellous origins for the family use." Ink recipes were never isolated in technical manuals; they were contained in household companion books. Ink and medicine were often linked: the Latin word *pigmentum* was used for both pigment and drug. Mitchell quotes from a book of family recipes that, along with several recipes for ink, "deals with all kinds of things good and bad, from recipes for apple pastries to cures for the King's evil," which is, of course, syphilis. In Act One of Rossini's *Barber of Seville*, Dr. Bartolo catches Rosina with ink on her fingers and suspects she's been writing to a rival:

Bartolo: What is the meaning of your ink-stained finger?

Rosina: Stained? Oh! Nothing. I burned myself and I used
the ink as medicine.

Ink was often used as an astringent, to treat burns. It was also used
to treat hemorrhoids, used as a dye for hair, and recommended as
a cure for baldness.

I wanted to make ink with my senior Book History class—we
could split up the work of grinding and try different solutions. (I
told them that ink had been an external mark of the literary: slave
traders were accused of fraud for splashing ink on a slave's clothes
to give the impression he could read and write. Jo in *Little Women*
has her "scribbling suit … a black pinafore on which she could
wipe her pen at will," which signifies to her family that she's in the
writing "vortex." So if we got a little ink on ourselves we'd be part
of a long tradition.) But where would I get gallnuts, gum arabic,
and ferrous sulphate? "Try Kremer supplies," Watson said. "It's all
online." Sure enough, it was. But how much to order? I didn't know
what a gallnut weighed, so I ordered a kilo of them just to be sure
I'd have enough for twenty students.

I took the materials into class: a bag of orange lumps that looked
like amber (gum arabic, the "binder"), a bottle of powdered green
stuff (the iron sulphate), a big bag of things that look like grey peas
(gallnuts, the "pigment"), and a selection of liquids (the "vehicle").
Three of the students had brought mortars and pestles. I brought
the hammers. I also brought a kitchen scale that measured ounces;
I said I wasn't about to buy one of those expensive scales that weigh
in tenths of grams. Who would buy those things? "*I* have one," said
Ashley, along with half a dozen other students in the class. "This
was cottage science," I told them. Tonight we would eyeball it.

You could feel the division in the class immediately. There
were those who embraced the casual aesthetic of the project, and

those who thought the whole enterprise was already slipshod and compromised. I separated them into four groups, hoping to get a combination of enthusiasts and skeptics in each. Looking at the recipe again in the classroom, on the brink of execution, I realized that even for home consumption the amounts were absurd. Eight ounces of liquid it called for. That's 236 millilitres, the size of the small bottle of beer I'd brought. Obviously a recipe for the days before the ballpoint. Ink could go mouldy and acidic; with four batches we'd still have more than enough for everyone to take home a sample.

"All right, class, we're adjusting the proportions here. Cut the liquid and everything else in half," I said, although this would make everything harder to measure on my crude scale.

"Okay," I continued, "weigh your gallnuts and put them in a bag and crush them." This at least I had thought out. But of course the hammers split the bag and scattered gallnut dust over the tables. "Okay, class, double up the plastic bags, and don't hit the gallnuts too hard." I'd also underestimated the density of gallnuts: my one-kilo bag was a hundred-year supply.

They carefully bashed the gum arabic (it was brittle and shattered easily) and ground the shards of gallnut and gum arabic together with the pestle. Group B under Jason was working like a Tour de France team, taking turns grinding in swift hard strokes.

"It's okay," I said, "it doesn't have to be that fine." They knew I didn't really know.

"We want to get these lumps out, Dr. Bishop," they said, not looking up as they pointed to the tiny motes and kept on grinding. I was pretty sure everything would dissolve anyway.

"Okay, now ladle in a little ferrous sulphate."

"How much?"

"Just a titch. Half an ounce."

"What's that?"

"Um, about half a flat tablespoon."

I'd read that there were four liquids you could use: water, vinegar, white wine, and beer. Since I was using a European recipe, I decided we should use a good European beer; I'd brought a bottle of Grolsch. "Okay, measure off four ounces and pour in the liquid." Everything went fine with the wine, the water, and the vinegar. Not so with the beer. Group C popped the cap and poured it in—and it came foaming back out. The mixture frothed up like a black milkshake.

"Dr. Bishop! It's going over!"

"It's okay. I brought paper towels."

Damn, none of the recipes said anything about letting the beer go flat.

Later I would encounter the term *flocculation* with sad recognition: it's a situation where a solute comes out of a solution in the form of flakes, or in colloids, where fine particulates clump together in *floc*. The adjective was more evocative: *flocculent*—resembling loose woolly masses. That was our beer ink. A floccing mess. The water and white wine both worked well, but Group A with the vinegar came up the winners: their ink, though it smelled a bit funny, had a lovely sheen to it—and the vinegar would arrest the formation of mould.

The crucial agent is the ferrous sulphate. This gorgeous green substance was an all-purpose tonic: it was prescribed for iron-deficiency anemia, and given indiscriminately to slaves in the eighteenth and nineteenth centuries for various ailments, though it did cause violent vomiting; more recently it was used as a flocculent in the purification of water, as a moss killer and lawn conditioner, and to add iron to Cheetos. In our iron-gall ink the darkening of the fluid came about through the gradual oxidation of the iron ions from ferrous to ferric (Fe^2 to Fe^3), which is why it had to be stored in airtight bottles. An excess of ferrous ions would create a rusty halo and ultimately cause the paper to disintegrate. On vellum (which is leather), the ferric ions then reacted with the tannin to form permanent chemical bonds with the protein in the leather—the same processes underlying tanning. With paper the ink made a mechanical rather than a chemical bond, penetrating the spaces between the cellulose fibres and becoming entangled with them.

In Tom Wharton's novel *Salamander* Nicholas Flood searches for a forbidden ink, the one that Johannes Trithemius claimed had the same chemical composition as fallen angels' blood, the one used to summon mournful, tormented spirits. The recipe begins, "Gather oak-galls, and grind minutely … add vinegar …." It sounded familiar. We didn't summon up tormented spirits (although with the spills and the smell we almost brought down campus security), but as we tried out our new old ink we knew we were making a connection with the past. Virginia Woolf, who wrote her novels with a straight pen, had mused in her diary, "I think it is true that one gains a certain hold on sausage & haddock by writing them down." My students who put aside their screens for an hour to write in longhand agreed. I handed out strips of acid-free artists' paper, and the students signed their names. "This will last two thousand years, right Dr. Bishop?" Right.

A CAVEMAN draws on a wall with a piece of charred firewood, complains that the flakes won't stick. His wife, distracted by his endless retelling of the hunt, has boiled down the mastodon bones too long, and tells him to dab his stick into the glop at the bottom of the pot. This works. Charcoal with a binder. But it's a little too gluey. She adds some water to make the stuff flow better. Now the liquid has all it needs: pigment, binder, and vehicle. Ink. The river of our cultural memory, the mark of our personal identity. One does gain a certain hold by writing it down.

A Note on Sources

INTRODUCTION

This is not a complete list of works cited or consulted, but it documents the main underpinnings of the book and is designed to give the reader avenues for further exploration.

Early on, when I was still thinking of this book solely as a commodity biography, I was inspired by Arjun Appadurai's *The Social Life of Things: Commodities in Cultural Perspective* (1986). In his introduction Appadurai argues that things have no meanings apart from those that human transactions endow them with—"their meanings are inscribed in their forms, their uses, their trajectories." The aim of my project was to trace the trajectory of ink within and across societies, from books to bodies, from commodities to sacred objects. There of course proved to be more trajectories than I anticipated.

I also found Bill Brown's distinction between the *thing* and the *object* useful in thinking about ink. In his introduction to *Things* (2004) he writes, "As they circulate through our lives, we look *through* objects.... A *thing*, in contrast, can hardly function as a window. We begin to confront the thingness of objects when they stop working for us: when the drill breaks, when the car stalls...." I discovered that ink slides from object to thing and back again, sometimes straddles the divide, and can function as a commodity in both realms at once.

Thorstein Veblen talks about candles in chapter six of *The Theory of the Leisure Class: An Economic Study of Institutions* (1899), a book that often feels startlingly contemporary. The passage from *Riding with Rilke* is from chapter three, "Archival Jolt."

PART I THE CRAFT OF INK

The Ballpoint

On Bíró and Goy, György Moldova's *The Never Ending Line: The Ball-Point Legend*, trans. David Robert Evans (ICO pic, 2001), is now available in a new edition with the title *Ballpoint: A Tale of Genius and Grit, Perilous Times, and the Invention that Changed the Way We Write*.

The doctor who committed suicide by tying a rope to a chair is mentioned in Frigyes Karinthy, *A Journey Round My Skull*, intro by Oliver Sacks, trans. Vernon Duckworth Baker (2008; 1939). Szép's description is quoted in Bob Dent, *Budapest: A Cultural and Literary History* (2007), and there is more on the city in John Lukacs, *Budapest 1900: A Historical Portrait of a City and Its Culture* (1988).

Lazlo Bíró's *Una Revolucion Silenciosa* (1969) remains available only in Spanish. I am grateful to Melba Montgomery for her draft translations that gave me access to the book and to Linda Gill, from the University of Texas at Austin, who provided the translations here.

The Borges edition that I was reading was *Brodie's Report: Including the Prose Fiction from "In Praise of Darkness,"* translated by Andrew Hurley (2005).

For Milton Reynolds I've drawn on the mass of material in the Special Collections Library at the University of Washington, Seattle, in particular the three biographies: Robert Leonard Rosenberg's PhD thesis, "The Ventures and Adventures of an Errant Entrepreneur, Milton (Ball-point) Reynolds"; "My Pen in Hand"—the autobiography of Milton Reynolds as told to Arnold Drake; and Gordon Schendel's massive typescript, "The Lucky Millionaire: A Biography of Milton Reynolds, 'Last of the Great Adventurers,'" which records the letters from outraged customers.

Reynolds makes things easy for a researcher, having collected his press clippings—though most of these are laudatory and only a few hint at the lawsuits. The quotations from *TIME* magazine, the accounts in Canadian newspapers of his 1947 round-the-world flight, and the copy of *Master Detective* (May 1950) with the story by W.T. Brannon that tells of the Reynolds's exploits in China are all in his scrapbooks and press files. More sober accounts are in Galen Rowell's "On and Around Anyemaqen," *American Alpine Journal* (1982), and Shiwei Chen's meticulously researched article, "The Making of a Dream: The Sino-American Expedition to Mount Amne Machin in 1948," *Modern Asian Studies* (2003).

Milton Reynolds's own breathless account of his travels, *Hasta La Vista,* sketches by Josie St. Hill (1944), republished with *Rocketing Round the World* (1947) is balanced by Thomas Whiteside's classic *New Yorker* article, "Where Are They Now? The Amphibious Pen" (1951).

For information on Marcel Bich I have relied primarily on Laurence Bich's fascinating *Le Baron Bich: Un Homme De Pointe* (2001). The translations are by Caroline Krzakowski, and in a few cases by Lloyd Bishop and myself; I of course take responsibility for any errors.

Printers' Ink
One of the best capsule histories of printing is in my favourite book on typography—poet and printer Robert Bringhurst's *Elements of Typographic Style* (1992, 2012). It is available online but the physical book is beautiful and worth buying.

Philip Ruxton's *Printing Inks, Their composition, Properties and Manufacture* (1918) and C. (Colin) H. Bloy's *A History of Printing Ink, Balls and Rollers 1440–1850* (1967) are more difficult to come by, but worth the trouble, as is Jim Stroud's unpublished monograph "Inks on Manuscripts and Documents," produced at UT Austin's Harry Ransom

Humanities Research Center. It is a concise and encyclopedic compilation of information about ink for conservators.

Stroud draws on C. Ainsworth Mitchell's *Inks: Their Composition and Manufacture including Methods of Examination and a full list of English Patents* (1916), which is available online, and so is *The Craftsman's Handbook "Il Libro dell' Arte" Cennino d'Andrea Cennini*, which is also still in print in an inexpensive edition.

The indispensable Joseph Moxon, *Mechanick Exercises on the whole Art of Printing* (1683–4) can now be downloaded to your phone, and Charles Manby Smith, *The Working Man's Way in the World, Being the Autobiography of a Journeyman Printer*, originally serialized in *Tait's Edinburgh Magazine* (March 1851–May 1852), a literary and political journal "of radical character" is also available in free online archives.

T.C. Hansard is not yet available for free, but his *Typographia: an Historical Sketch of the Origin and Progress of the Art of Printing; with Practical Directions for Conducting Every Department in an Office: with a Description of Stereotype and Lithography, illustrated by Engravings, Biographical Notices, and Portraits* (1825), is accessible in reprint volumes.

Anything by Robert Darnton is worth reading—scrupulous scholarship written with journalistic flair—see *The Great Cat Massacre and Other Episodes in French Cultural History* (1985). Philip Ball is another scholar worth reading for his writing—see *Bright Earth: Art and the Invention of Color* (2001). Less eloquent but still fun to dip into is *Henley's Twentieth Century Book of Formulas, Processes and Trade Secrets*, edited by Gardner D. Hiscox, M.E., New Revised and Improved Edition (1934), available online.

PART II THE ART OF INK

For background on the wider cultural context I am indebted to Jean Francois Billeter, from *The Chinese Art of Writing* in Rothenberg and Clay, *A Book of the Book*: *some works & projections about the book & writing* (1990); Nigel Cameron, *The Chinese Scholar's Desk* (2003); Craig Clunas

Superfluous Things: Material Culture and Social Status in Early Modern China (1991); Joseph P. McDermott, *A Social History of the Chinese Book: Books and Literati Culture in Late Imperial China* (2006). Research also requires that one read erotic works, such as *The Golden Lotus*, trans. Clement Egerton from the Chinese of Chin P'ing Mei (1939).

Simon Winchester's *The River at the Centre of the World: A Journey Up the Yangstze, and Back in Chinese Time* (1998) is curiously Anglocentric at moments but the descriptions of the Yangstze are superb. Peter Hessler's *River Town: Two Years on the Yangtze* (2001) brilliantly renders the process of his gradual, and always partial, integration into the society of a Chinese city. Ma Jian's *Red Dust: A Path Through China*, trans. Flora Drew (2001) offers the perspective of someone alienated in his own country.

A good way into Chinese calligraphy is Chiang Yee's book, which began as a series of lectures: *Chinese Calligraphy: An Introduction to Its Aesthetic and Technique*, with a Foreword by Sir Herbert Read (1938). A more recent authority is Tseng Yuho, *History of Calligraphy* (1993). For Dong Qichang (and note the various spellings of his name), see the sumptuous collection in two volumes edited by Wai-Kam Ho, *The Century of Tung Chi'i-ch'ang* (1992).

Amy Tan, *The Bonesetter's Daughter* (2001) takes you into the culture of ink, and if you want an authoritative account of ink-making turn to Joseph Needham's *Science and Civilization in China*, specifically *Volume 5, Chemistry and Chemical Technology, Part 1: Paper and Printing by Tsien Tsuen-Hsuin* (1985). Tsien is responsible for the ink section and he also published *Written on Bamboo and Silk* (1962). Needham is an intriguing eccentric, caught by Simon Winchester in *The Man Who Loved China* (2008).

Wang Chi-chen's "Notes on Chinese Ink." *Metropolitan Museum Studies* (1930) provides an excellent introduction to the subject of inksticks, which is developed in innumerable articles, such as R.H.

van Gulik, "A Note on Ink Cakes," *Monumenta Nipponica* (1955); Lily Chia-jen Kecskés, "Art and Connoisseurship of the Ink," *Ars Decorativa* (1984), and Anita Siu, "Ink-making in China," *Orientations* (June 1997).

If you want to pursue ink down to the level of electron microscopy, and to understand the shift from pine soot to lampblack (and who wouldn't?) see Joseph R. Swider, Vincent A. Hackley, John Winter, "Characterization of Chinese ink in size and surface," *Journal of Cultural Heritage* (2003), as well as Winter's earlier article "Preliminary Investigations on Chinese Ink in Far Eastern Paintings," in *Advances in Chemistry Series*, ed. Robert. F. Gould (American Chemical Society, 1974). With pictures.

On the Ming Dynasty I have benefited from Timothy Brook, *The Confusions of Pleasure: Commerce and Culture in Ming China* (1998); James Cahill, *The Distant Mountains: Chinese Paintings of the Late Ming Dynasty, 1570–1644* (1982); and the indispensable Jonathan D. Spence. I have drawn specifically on *The Memory Palace of Matteo Ricci* (1984) and *Return to Dragon Mountain: Memories of a Late Ming Man* (2007), but I have consulted other works by Spence, such as the magisterial *The Search for Modern China* (Norton, 1990) and the writing is always compelling.

On Cheng and Fang, moving far beyond Wang's article mentioned above, and even Sewall Oertling's essay "Patronage in Anhui During the Wan-li Period," in *Artists and Patrons: some social and economic aspects of Chinese paintings* (1989), Lin Li-chiang's 1998 Princeton PhD dissertation, "The Proliferation of Images: The Ink-stick designs and the printing of the *Fang-shih Mo-P'u* and the *Ch'eng-Shih Mo-Yüä*" offers a detailed examination of the catalogues and an analysis of their cultural importance. The prose is tough sledding but the research invaluable.

As an introduction to Chinese poetry, at a bed and breakfast on Vancouver Island I came upon (and stole) Greg Whincup's *The Heart of Chinese Poetry* (1987), which I recommend highly. I've also consulted *The Columbia Book of Later Chinese Poetry, Yüan, Ming, and Ch'ing Dynasties*

(1297–1911), trans. and edited by Jonathan Chaves (1986), and various online sites. My thanks to Leilei Chen and Xue Sheng for tweaking some of the translations. My favourite Du Fu volume is *Du Fu, A Life in Poetry*, trans. David Young (2008), but I read him in conjunction with Rewei Alley's *Du Fu Selected Poems* (2011), David Hinton's *The Selected Poems of Tu Fu* (1988), and *Spring in the Ruined City: Selected Poems of Du Fu*, translated by Jonathan Waley with calligraphy by Kaili Fu (2008).

PART III THE SPIRIT OF INK

I found useful background in Albert Hourani's *A History of the Arab Peoples* (1991) and Karen Armstrong's *Muhammad: A Prophet For Our Time* (2006). I also frequently consulted the *Encyclopedia of Islam* in both its print and online editions. In Uzbekistan I relied on Calum MacLeod and Bradley Mayhew's fine guidebook, *Uzbekistan: the Golden Road to Samarkand* (2008). The image of the Persian carpets hung up to the sky is stolen from Robert D. Kaplan, *The Ends of the Earth: A Journey to the Frontiers of Anarchy* (1996) in which he points out that the sand of the Registan ("place of sand") was used to soak up the blood from public executions. Orhan Pamuk's *My Name is Red*, trans. from the Turkish by Erdag M. Goknar (2001) frustrates some readers with his long sentences, but he brings the archives to life.

Alexander Burnes, *Travels into Bokhara: being the account of a journey from India to Cabool, Tartary, and Persia; also, Narrative of a voyage on the Indus, from the sea to Lahore, with presents from the king of Great Britain; performed under the orders of the supreme government of India, in the years 1831, 1832, and 1833* (1834) is available online and worth dipping into if only for the smirking frontispiece portrait of Burnes.

For background on the calligraphy I turned to Abdelkebir Khatibi and Mohammed Sijelmassi's lavishly illustrated *The Splendor of Islamic Calligraphy* (1976), Johannes Pedersen's *The Arabic Book*, trans. Geoffrey

French (1946, 1984), and Anne Marie Schimmel's *Calligraphy and Islamic Culture* (1984), as well as her article and others in George Atiyeh's collection of essays, *The Book in the Islamic World: The Written Word and Communication in the Middle East* (1995).

The translation of Al-Mu'izz ibn Badis, "Staff of the Scribes and Implements of the Discerning With a Description of the Line, the Pens, Soot Inks, *Liq*, Gall inks, Dyeing, and Details of Bookbinding" (circa 1025, from a manuscript in the Oriental Institute, University of Chicago) is contained in a quirky article by Martin Levey, "Mediaeval Arabic Bookmaking and Its Relation to Early Chemistry and Pharmacology," *Transactions of the American Philosophical Society* (1962).

For gallnuts I consulted C. Ainsworth Mitchell's *Inks* and David Carvalho's *Forty Centuries of Ink*, which has an encyclopedic subtitle that rivals those of eighteenth-century novels: *A Chronological Narrative Concerning Ink and Its Backgrounds, Introducing incidental observations and deductions, parallels of time and color phenomena, bibliography, chemistry, poetical effusions, citations, anecdotes and curiosa together with some evidence respecting the evanescent character of most inks of to-day and an epitome of chemico-legal ink* (1904). It is available online but if you want to find your way around in it you should order the print-on-demand version.

For cochineal and all things red see Amy Butler Greenfield's fascinating book *A Perfect Red: Empire, Espionage, and the Quest for the Colour of Desire* (2005). As for the PETA website, it seems to be less pithy than when I first accessed it. It now says, "Red pigment from the crushed female cochineal insect. Reportedly, 70,000 beetles must be killed to produce one pound of this red dye."

Shirin Neshat's photographs have been published in *Women of Allah* (1997). The passages from Ayaan Hirsi Ali's *Infidel* (2007) are from chapters 2 and 5. The film "Submission" is available online and the subsequent events have been explored by Ian Buruma in *Murder in Amsterdam: The Death of Theo Van Gogh and the Limits of Tolerance* (2006). My discussion of translations of the Qur'an derives from chapter three of Reza Aslan's *No god but God: The origins, Evolution, and Future of Islam* (2005).

On the Celestial Pen I learned from Uzma Mirza, "*The pen & The Inkpot: A Muslim woman's spiritual Art, through the Science (ilm), of knowing the heart (Qalb)*" (online), and Luce López-Baralt's article trans. Marikay McCabe, "The Supreme Pen (Al-Qalam Al A'lā) of Cide Hamete Benengeli in *Don Quixote*," *Journal of Medieval and Early Modern Studies* (2000).

PART IV THE RENAISSANCE OF INK

On hydrodynamics, the *Pen World* article is from August 2012, and for the scientific article see Jungchul Kim, Myoung-Woon Moon, Kwang-Ryeol Lee, L. Mahadevan, Ho-Young Kim, "Hydrodynamics of Writing with Ink," *Physical Review Letters* (2011), and on haptics see Anne Mangen and Jean-Luc Velay (University of Stavanger, Norway; Mediterranean Institute for Cognitive Neuroscience, Marseille) "Digitizing literacy: reflections on the haptics of writing," in *Advances in Haptics* (Intech, 2010).

The debate about cursive writing is easily pursued online: the reader can google "Jack Lew's signature"; Andrew Coyne, "Losing longhand breaks link to the past"; Lee Rourke, "Why creative writing is better with a pen," and other articles about the end of writing. Daniel Chandler's seminal article, "The Phenomenology of Writing by Hand," *Intelligent Tutoring Media* (1992) is available online.

The quotation from Robert Pirsig is from the second page of *Zen and the Art of Motorcycle Maintenance: An Inquiry Into Values*; the quotation from Steven Alford is from his conference paper, "Popular Travel Narratives and the Motorcyclist: Traveling In, Not Traveling Through," English Language and Literature Association of Korea (2011). The Roland Barthes quotation is from his 27 September 1973 interview in *Le Monde*, translated by Linda Coverdale as "An Almost Obsessive Relation to Writing Instruments" in *The Grain of the Voice* (2009). Anne Fadiman's essay is in *Ex Libris: Confessions of a Common Reader* (1998). "Calamophile" is the blog of one Katherine Bishop (no relation).

Molly Bloom writes in the "Ithaca" episode of James Joyce's *Ulysses* (1922), Jo scribbles in chapter twenty-seven of Louisa May Alcott, *Little Women* (1868), and the magical ink appears in "The Well of Stories" section of Tom Wharton's *Salamander* (2001). Virginia Woolf's reference to "sausage & haddock" is the second-last entry in her diary, 8 March 1941.

Acknowledgments

I want to thank my wife, Hsing, for her faith in the project, my toddler, Thomas, for his endless wonder, and my adult children, James and Erin, for their encouragement.

My thanks to Maria Scala for her careful reading of an early draft of the book. For the opportunity to present material from the work in progress, I want to thank Dr. Susan Hamilton for inviting me to deliver the Broadus Lectures at the University of Alberta; Dr. Dave Buchanan for inviting me to speak on ink and identity at MacEwan University; Dr. Michael Epp for inviting me to speak at Trent University; Dr. Leonard Diepeveen for inviting me to Dalhousie University; Lynn Coady and Marina Endicott for including me in one of their Literary Saloons; Ian Brown and Robyn Read for bringing me to the Banff Centre at different times; and Wade Kelly of Nerd Nite Edmonton for promoting scholarship with alcohol and endorsing the title, "Crush My Gall Nuts Baby and I'll Stay With You Till the End of Time."

My thanks to Denise Teece for her generosity in allowing me to look at the original paintings by Bihzad at the Metropolitan

Museum of Art, New York; Jim Whittome at the Mactaggart Collection, University of Alberta, Edmonton; and to Gary Lundell at Special Collections, University of Washington, Seattle, for tracking down images. I'm grateful to Jeannine Green, Robert Desmarais, and Jeff Papineau at Special Collections, University of Alberta, for their help at various stages of the project.

I wish to acknowledge the support of the Social Sciences and Humanities Research Council for a grant that made much of the travel possible, and the University of Alberta for the sabbatical that made the writing possible.

In addition to those mentioned in the book, I want to thank the individuals who helped with the work in many different ways, from dinners, to pints, to research: Miki Andrejevic, Natalie Anton, Norman Bishop, Paul Byrne, Sahar Charradi, David Cheoros, Stephan Christianson, Brian Evans, Brianna Erban, Judy Gatto, James Gifford, Tabitha Gillman, Ross Jopling, Myrna Kostash, Jim Marggraff, Alex McGuckin, Jill McIntosh, Peter Midgely, He Ning, Lindsay Parker-Gifford, Karen Platten, Andreas Schroeder, Aida Yared, Karen Wun, Kornel Zipernovsky. I know I'm missing some, but thank you all.

Finally, I wish to thank publisher Nicole Winstanley for believing in the project, copy editor Karen Alliston for untangling syntactical knots, and the production team at Penguin for creating a handsome volume. Above all I want to thank my excellent editor Justin Stoller for helping to find a line out of what had become a morass of ink.

Index